To John,
With best wish[es]

Malcolm, Jane and Richard.

PORTRAIT OF THE POTTERIES

Portrait of
THE POTTERIES

W. A. MORLAND

Photographs by the author

ROBERT HALE LIMITED

© W. A. Morland 1978

First published in Great Britain 1978

ISBN 0 7091 6573 0

Robert Hale Limited
Clerkenwell House
Clerkenwell Green
London EC1R 0HT

FILMSET BY WEATHERBY WOOLNOUGH, NORTHANTS,
PRINTED IN GREAT BRITAIN BY
LOWE & BRYDONE LTD., THETFORD.

CONTENTS

ILLUSTRATIONS

I

BURSLEM

HEMMED in for centuries by their hills, the Potteries villages early developed a healthy tradition of independence and self-reliance. They were much more a part of the bleak, northern moors than the lush meadows of the Midlands. As they grew in prosperity and importance to become, as a federation, amongst the first dozen towns in Britain in size, they retained both their moorland stoicism and more than their fair share of that peculiar parochialism, always bred of neglectful isolation. Each was to prove a blessing and a curse.

The first hint of this unique local character is that, on first visiting the place, the stranger is immediately confused by his inability to find a centre that seems to do justice to what he knows to be a large industrial town, dignified by the title 'city' and possessing a Lord Mayor. Such august personages are normally enshrined in a town hall—at least for a year—and there is no shortage of town halls in which to enshrine them. There is a town hall every couple of miles, and each would do justice to a small town like Aylesbury. None of them is outstanding in the usual sense, but each has its good points and is very like the last in character if not in style. Each embodies something of the sturdiness of the Potteries, and has its surrounding shopping area scattered about it in a homely muddle. Each is now, regrettably, acquiring a sterile and desperate collection of concrete cornflake boxes and toy-town sets of stairs and tunnels, cringing under the title 'precinct', a name more suited to the macabre activities of New World law enforcement agencies than the shopping needs of the quiet, friendly people of the Six Towns.

However, the existence of the Town Halls helps to explain

"

why local folks refer to the area as "the Potteries"—very much in the plural—and not as "Stoke-on-Trent". And why, when you ask the inevitable Midlands question "Where do you come from?" they answer "Burslem" or "Hanley" or "Stoke" or "Longton" or "Tunstall" or even, surprisingly, "Fenton" because almost everyone forgets Fenton and that is why we hear of the Five Towns.

Even the local motor vehicle registration letters deride Fenton—VT—'five towns', and no one has any doubts about which one is left out.

Arnold Bennett, the Potteries' chief chronicler, is usually blamed for the mistake in the arithmetic, but when the first Queen Elizabeth, in her infinite wisdom, imposed on the ancient parishes of the realm the statutory obligation to make provision for the relief of the poor of the immediate area, her commissioners divided the huge parish of Stoke, one of the largest and most ancient in Staffordshire, into five units, jumping the gun on Arnold Bennett by almost exactly three hundred years. In that instance the one left out was Burslem, the 'Mother of the Potteries', who, even at this early date, was making her mark as a local centre for potting, and already had a measure of independence, having achieved the dignity of her own tiny church in the reign of Henry VIII, the tower of which still stands to remind the townsfolk of their humble origins. Despite the efforts of the locally great to enhance the Tudor structure in the Georgian style, the church of John the Baptist, Burslem, remains what it always has been, the poor moorland church of a poor moorland village, not the grand Renaissance temple of a great industrial city, such as was built on the nearby Hill Top later on.

But that is all to the good, for Burslem, standing as Wesley said, "A scattered town on top of a hill", holds her skirts closely to avoid three of the brooks that do so much to dictate the once and future shape of the Potteries, and preserves more buildings of the Georgian period than the rest of her neighbours put together.

Not that 'neighbours' is a particularly good description of reality, since most skilled urban observers and visitors of cities would expect to find the city an expansion of a central

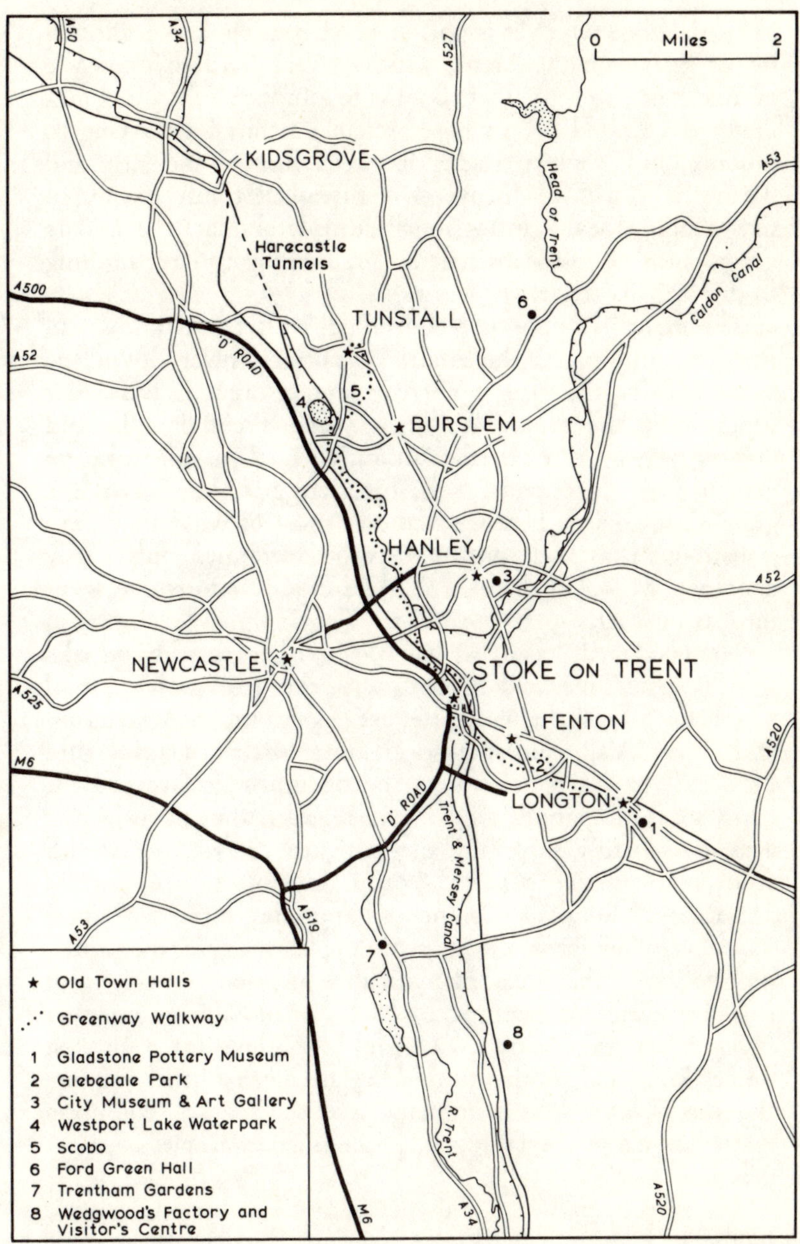

The A500, 'D' Road was renamed 'Queensway' in 1977, in honour of the Silver Jubilee.

industrial zone which it is not, or at least to be visibly affected by, or sitting on the Trent, which is very hard to find. We cannot turn to our text books to understand this four-hundred-year-old lady, whose present portrait I am trying to produce, and whose character has been formed anciently and deviously. What we discover is a thread of thinly connected towns about seven miles long, completely lacking a true geographical or historical centre, and seeming to have nothing to do with its river.

The infant Trent is not concerned with the old town of Burslem, and enters the affairs of potters lower down the valley, where the city is narrow, and escape to the easier meadows of the county is speedier. The Trent dawdles and dodges its way through the unforgiving streets of Stoke, to give little but its name. Still, for that gift the Towns are grateful, since little here is given and most of what they have is hard-won, as their principal export, ceramics, only really exists as a symbol of the many skills and joint efforts that went into its making.

The good, red, local clay has been used for bricks and tiles since Roman times and beyond, with the timber of the local woods for fuel. The ancient sites are easy to find at Chesterton and Trent Vale and the local museums at Hanley and Newcastle have superb exhibits and documents relating to the many local excavations which have revealed the efforts of past ages. It is hardly surprising that an area deriving its wealth from holes in the ground should develop a rare skill in archaeology in all its branches, and the numerous local societies do not have the tidy middle-class appearance common to some other areas; many of their members emerge from marl-holes and coal-pits, scrub and eat and descend smaller holes exchanging their excavators and picks for a broken dinner knife and tooth-brush. They need very little training, the sum of local skills is immense, and the rate of demolition in the service of the false god, progress gives ample scope for practice.

The clay upon which the ancient potters thrived is found in the rocks of the Black-Band group; this is a remarkable stratum containing iron, coal, limestone and various clays including fire-clay. In some places the iron-coal-limestone mix was such that piles of the rock, arranged like an open beehive and ignited, pretty well refined the iron ore which made it very easy to smelt.

This unique geological marvel gave rise to an early iron industry in many North-Staffordshire towns, which was very handy for both potting and farming, since you cannot have an industry in modern terms without iron to back you up.

Throughout the Middle Ages, settlements in our valleys and hills were what you would expect from a normal reaction to the land and the weather. Burslem was an early hill-top settlement—the town was called Barcardeslim in the Domesday Book, and Burgheard's Lyme by the Anglo-Saxons. It is not surprising that the Domesday scribes made a mess of that and so many other names; just imagine what the response of the twentieth-century Anglo-Saxons would have been had Hitler's storm-troopers tried to make an inventory of similar proportions in 1940. Grudging is hardly the word. In the days of William the Bastard we have the hilarious situation where truculent Saxons, recently deprived of a great victory, are questioned by Gallic-speaking scribes who had to translate the mumbled falsehoods of a reluctant peasantry from a variety of carefully nurtured dialects into Latin shorthand on calf-skin with a feather dipped in a mixture of soot and bulls' urine.

However, the name means Burgheard's Elms or Woods, and although there is nothing now to be seen of the forest that attracted him fifteen hundred years or so ago, there is one small town that is glad he stopped there on his family's incredible journey of centuries through thousands of miles of the woods and lakes of northern Europe. In Burslem, windy farmsteads had the usual potting activity found in areas where the clay was suitable for the coarse medieval pottery, and there was enough wood for fuel. As well as the bounty of local geology, nearby Derbyshire had plenty of lead ore available for sealing and decorating the 'ware' and monastic settlements like Hulton Abbey to the east made use of 'pit-coals' as early as the fifteenth century and there were mines of a sort in the 1280s.

Although it is not entirely valid to assume that an area devoted to an industry depending on minerals might have developed along a particular group of minerals that could be got at easily, it is true that in the early stages of the pottery industry local clay was useful and local fuel was important. The same could be said for a good many other places. We have to cast around a little more widely to account for the present existence of the Potteries.

At the end of the seventeenth century some spectacular changes in agriculture occurred which did not suit either the heavy clays of North Staffordshire nor the small-holders of Burslem, and an increasing number helped out their family budgets by making and selling pots from the clay that hindered their ploughs. Many of them made butter pots for the butter market at Uttoxeter. This doubtful delicacy must have reached the capital in a highly odorous condition, but the Londoners certainly liked it or the trade could not have been so widespread. By 1661 the Burslem butter-pot makers were sufficiently flourishing to make the government of Charles II sit up and take notice. In 1662 an Act of Parliament called upon them to restrict the weight of their pots to about six pounds, so that a twenty-pound load could contain a stone of butter.

What they were up to was this: they made the pots with great, thick bottoms, like a modern cold cream jar, and they made them anonymously, which was very bright. The weight stipulation was obviously, therefore, aimed at the potters, but it seems that the butter sellers were not above suspicion. To quote Dr Plot, who, I must admit, has often been considered to be more entertaining than accurate:

> According to an Act of Parliament made about 14 or 16 years agoe, for regulateing the abuses of this trade, in the make of the Pots, and false packing of the butter; which before sometimes was layed good for a little depth at the top, and bad at the bottom; and sometimes set in rolls only touching at the top, and standing hollow below at a great distance from the sides of the pot: to prevent these little country moorelandish cheats (than whom no people whatever are esteemed more subtile) the factors keep a surveyor all the summer here.

That took care of the brethren from Uttoxeter and another

little idea from the Act should have sewn up the hardware end:

> Every potter shall set upon every pot which he shall sell for the packing up of butter the just weight which shall be of every pot when it is burnt together with the first letter of his Christian name and his surname at length upon pain for every pot of one shilling.

This all sets us a splendid example of law-keeping. First, no one has ever been able to explain how the rough and ready technology of the day would possibly provide the potter with a sure knowledge of even whether his pots would come out all right, let alone guess accurately what they might weigh. Secondly, very few butter pots marked as ordered have ever been found. I suspect that these pioneers of packaging thumbed their noses at impracticable laws from their splendid isolation in the Pennine foothills, and a complaint made by Daniel Everard of Uttoxeter at Stafford in 1682 seems to confirm this. Briefly he bitterly bemoans the fact that two "earthen potters", Thomas Daniell senr and Tho. Cartwright, formed a monopoly of the making and supply of butter pots in Burslem, Hanley Green and Stoke. They almost doubled the price, according to poor Mr Everard, "contrary to law and to the great prejudice of this County, and to the Evill example of others".

We cannot really blame the potters for attempting to stabilise the price of an article that must have been dirt-cheap in the first place, and upon which such a heavy penalty was attached in case of non-observance of a law that was 50 per cent impossible anyway.

That was not the first brush the Staffordshire potters had with the law. It is necessary to know that you cannot just dig up clay from the garden and make it into a jasper ashtray; certain rites and mysterious operations have to be complied with, and the first of these is known as 'levigating' which is a rather refined way of saying 'picking the lumps out'. Now this cleaning and mixing of clay can be a hard job and is absolutely essential if ruinous explosions in the kiln—which is where pots are baked or fired—are to be avoided. Treading it in the bath like grapes is one way, and it is supposed to be very good for the complexion. To do the job on a large scale one either needs the help of a great number of family and friends or a herd of cattle. This is where the first known local potters, belonging, appropriately, to the Adams family, fell foul of the law. They noticed that the passage of many feet, human and otherwise, had turned the clay of the 'road' between Burslem and Sneyd into a first class batch of pot-clay. Being very bright and enterprising they proceeded to dig up the clay and take it off to their nearby pot works for further treatment. Not surprisingly the hole was noticed and Richard Adams and his brother William were fined for 'digging clay by the road'. The year was 1448. The Adams family are always cropping-up in Potteries history, and made an enormous contribution to the area in many ways. In 1563 a Thomas Adams of Burslem left his "best yron chymney" to his son William and his other chimney to his daughter Ellen. One assumes that the chimneys were the kilns. This same family claims the distinction of inheriting the potting rights of the monks of Hulton Abbey when Henry VIII dissolved it, and the firm is still in business today.

Those Daniels who were busy engrossing butter pots in 1682 were from a long line of potters, too. They were making pottery in the area in the days of Good Queen Bess, and some of the butter-pot fragments that did get marked as prescribed were dug up with the Daniels' name on them near Burslem Town Hall.

Returning to Dr Plot's view of the area, which was published in 1686, he found that the "fabrication of Common Vessels" continued in a very rude state, yet he was most respectful concerning Burslem, as having the largest potteries then known to exist. The controversy over when Dr Plot actually visited Staffordshire, so that the processes he describes can be dated, it easily resolved. He certainly was not here in 1686, the year of his book's publication, as he had been indulging in the author's perennial pastime of fobbing-off his publisher about a delivery date for his manuscript. He inserts a whining defence of his actions at the beginning in a preface which ends with the statement that he is not concerned to give certain persons satisfaction.

Since the Act was passed in 1662 and he says it was fourteen or sixteen years ago, we arrive at a date of either 1676 or 1678; either way he kept his publishers waiting for at least eight years, and we get a lively, if whimsical, view of our area in the 1670s.

Even Robert Plot divined that the area's advantage was largely geological, but he did not know why Tunstall, with a similar geology, did not have potteries as great as Burslem's, though he must have noticed that Burslem was on a road, which was all the advantage it needed.

Close to the town centre was good quality coal, clay suitable for the potters' wheel and saggar marl—clay for making the containers used to protect the pots during firing—all near to the surface. And in those early days local pots were in demand since, as he observes, "They draw them [from the kilns] for Sale, which is chiefly to the poor Crate-men, who carry them at their backs all over the Countrey."

Although the production of 'earthenware' flourished in Burslem for geological reasons, the imports of fine tableware from overseas, especially China and France, and the availability of whiter, much less coarse pots from the potteries of Lambeth and Fulham provided some fierce opposition for the local 'common vessels'.

Since they wanted to preserve their trade, Staffordshire potters made ceaseless experiments to improve their wares. The introduction of salt glazing in 1690 gave them the chance of producing an article of superior quality, and, paradoxically, they learned about the new process from the works of one of their chief competitors, John Dwight of Fulham. The Staffordshire potters were always on the move, seeking inspiration and new sources of material, and, naturally enough, indulging in every form of industrial espionage known to a scientific community only recently emancipated from the stigma of witchcraft.

What they saw delighted them. Pots were formed, put in the kiln to fire, and then, when the fire was really hot, shovelsful of salt were hurled into the 'bags' as they call the hearths in a pottery. The incandescent salt settled on the pots and when they were cool they had achieved a lovely glossy coat. Or so they thought. Unfortunately, the highly coloured Burslem clays tended to melt and bubble and collapse at the high temperatures needed for the new process.

Local potters were already importing the whiter clays from Dorset and Devon to hide the colour of local wares by lining the insides and decorating the outsides, and a mixture of these 'ball-clays', brought in in huge balls slung on the sides of pack-horses, with local clay enabled salt-glazing to be done.

The local lads continued to experiment and one of them, William Astbury of Shelton, is credited with the introduction of flint as a potters' material. The story is that William, stopping at a pub on his way home from London, noticed an ostler blowing something into a horse's eyes. Whether he was humanely interested in the horse's welfare, or just plain nosey, we do not know. He discovered that the 'medicine' was made from flints which had first been burned and then crushed, producing a fine, white powder. Now Billie Astbury was in the market for fine white powders to improve his pots, and he found that the addition of this 'calcined' flint to the clays they already had, both strengthened and whitened the product. This mixture was the basis of the fine white wares upon which the prosperity of North Staffordshire was built, and upon which, in a modified form, we still depend.

Once this breakthrough in technology had been made, improvements continued for many years, and as a result the villages expanded. This expansion followed the lanes linking the settlements. The lanes themselves were decided by one of the main features of the area, the Fowlea Brook.

The new 'white stone ware', now frequently glazed with a hideously poisonous mixture of clays and white lead, was universally popular, and by 1770, the demand was so great that the expansion of premises to make it, and the housing for the 'operatives', overflowed the old towns of Tunstall and Burslem which now bulged towards each other along the east side of the Fowlea valley, over the eastern ridge, and on towards Hanley where the coal was being dug. Manufacturing capacity outstripped the transportation facilities of the immediate area, since the early turnpike roads all missed the potteries. Such roads as there were, were hopelessly inadequate. In an England suffering all the trials and tribulations of an agricultural and industrial revolution the potteries were booming.

In the time of the German Georges the Fowl Hey valley was a peninsula of employment in a sea of shattered hamlets and burning thatch. Farm workers 'on the tramp' flooded in on every tide of land enclosure and agricultural improvement. Names now long associated with the potteries can be found on the tombstones and in the registers of churches from a radius of over thirty miles.

Down the banks and over the brooks from Newcastle and the Shropshire borders they came: more trudged up from the once hospitable fields of the south; yet more came scrambling over the hills from Derbyshire and the east. Each put a shoulder to the slowly moving wheel of new industry; each added something to the richness of Potteries life with accent, custom or tradition; until the whole, stewed together with hardship, slip and lung-searing dust, produced the Staffordshire Potter.

One cannot be concerned with the Potteries for long before the name Wedgwood crops up, and it was in the first period of expansion that two brothers, Thomas and John Wedgwood, turned the manufacture of pottery from a small-scale domestic industry into a large-scale commercial enterprise. On the corner of what is now Wedgwood Street and Moorland Road they built and operated a pot-works and from the very considerable profits they built the 'Big House'. The pot-works has long since been built over, but the Big House is still there in all its glory. It is far and away the finest building in the area to survive and successive owners have, with commendable taste and foresight, retained much of the interior as it was in 1751 when it was built.

Unfortunately, the same cannot be said of the nearby Red Lion Inn, where one used to be able to see the coal outcropping in the cellar, an interesting example of the benevolence of local geology. Since this is probably the oldest house remaining in Burslem, it is a relief to see on the front.

a tile dated 1675 with the initials R.D.S., amongst the Swiss-cottage mockery that now seems to be demanded of urban hostelries of any age.

These two significant buildings are watched over by an especially charming guardian angel perched decorously on top of the splendid old town hall in the square opposite. This town hall is real vintage. It must rank very high on anyone's list of favourite buildings. Its squat, both-feet-on-the-ground strength, with embellishments of most building styles you can name, is far too charming to be pretentious in the blind, uncompromising way that so many public buildings up and down the country manage to be. It took the builders three years to complete the devious designs of Mr G. T. Robinson of Wolverhampton, and when, in 1857, it was done it created a sensation; in fact, it still does, when anyone bothers to look at it. Where else can you see a Corinthian temple perched on top of a studiously rusticated railway bridge? The porch with its clock has now taken root so firmly it is difficult to know whether it was really an afterthought or part of the original aspiration. One thing is certain, the whole of the valley would be the worse for its removal, which must have been on the cards from time to time ever since the new town hall, which cowers away in a corner, completely over-awed by its elder parent, was built in 1911. Now, praise be, it seems to be safe, since a further use has been found for it. Having done service as a library for years, it plays a new part in enriching the quality of life of the Towns as a recreation centre. No doubt the aches and pains inflicted in the pursuit of physical fitness are compensated by the attractiveness of the building and the privilege of using the magnificent entrance.

Burslem's golden angel watches over the most fascinating concentration of assorted buildings and industries you could hope to find. Most of the land to the south of the Town Hall, right up to Queen Street, was once occupied by the famous Brickhouse Works. One of the oldest traceable potteries, it was in the hands of yet another member of the Adams family from 1657 to 1687, and produced some of the characteristic black mottled ware, examples of which are in the Ceramic Museum, and some of which was found on the site during excavations. The Adams went on producing ware there until 1762, when it had its most illustrious tenant, Josiah Wedgwood, who developed his 'Egyptian Black' and cream-ware there. This cream-coloured ware was admired by Queen Charlotte, amongst others, and it was named 'Queen's Ware' after her. She is the Queen in Queen Street, named in gratitude for her patronage, which put the Royal seal of approval on Staffordshire pots. It was to urge his workforce to punctuality to fulfil the orders on his bulging books that Josiah installed a bell on top of the old Brick House. The custom in other works at the time was to sound a horn when the hour was nigh. The bustling Mr Wedgwood—who was in immediate competition with a myriad of relatives—had little faith in his workers' ability to pick out and be urged on by one horn amongst so many, however much the bugler changed his tune. The answer was a bell. The result must have been quite startling in the sonorous, sulphurous six-o'clock-in-the-morning air of Burslem, because Brick House became Bell House Works', and work it certainly did. That bell was just one of the minute advantages that gave J. Wedgwood the edge on his rivals and led to his being the most successful and famous of all the Staffordshire Potters. This remarkable man had started as an apprentice to his brother Thomas, who had inherited the family business, and did not care to take his younger brother into partnership, which was the coming trend as the potteries changed from small, cottage concerns employing families to a factory base employing large numbers of strangers. Having achieved the status of master-potter, Josiah set up on his own at Cliff Bank in Stoke. His partner was Thomas Alders, a practical potter, who already worked at the Cliff Vale 'bank' with John Harrison who had 'bought-in' to the business. Josiah's small inheritance of twenty pounds—from his father—was injected into the business, which seems to have depended on his skill. That was in 1752; by 1754 Josiah had joined forces with one of the world's most distinguished potters, Thomas Whieldon of Fenton Low. This arrangement lasted for five years, and it was whilst working with Whieldon that Wedgwood developed the beautiful, though deadly, green glaze for which he became so famous. His later letters show how crowded was his brain with schemes of all sorts, ceramic, scientific and commercial. Junior partnership, with all its constraints, must have aggravated that sense of frustration at being unable to do enough, to which he so often referred in later life. So, in 1759 he rented the Ivy House works, from John and Thomas of the Big House, for ten pounds a year. The Ivy House was just across the street from the Big House, near the old town hall. The site has since been a market hall and is now partly park bench, partly band stand and partly public convenience.

When Josiah moved from the Ivy House in 1762 to the Brick House Pot Bank, he was really on his way to the top. Successful enough to be able to install yet another Thomas Wedgwood, a cousin, as manager of his "useful wares" concern in 1766 he started his lengthy campaign to persuade Thomas Bentley, a cultured and successful Liverpool business man, to join his growing empire. His order books were so full that his Burslem works could not cope with the orders, and the roads in and out were so bad they could not cope either. Josiah arranged in 1769 to join with James Brindley and others in the construction of a canal, the course of which was again largely dictated by the course of the Fowlea Brook, and which was designed to speed up the delivery of fine clays and flints from Devon and Dorset, and to take away the finished product.

Wedgwood chose to place his new factory near the canal on the Ridge House Estate for which he had paid £3,000, and opened it in 1769. Even this was not enough capacity and in 1780, Josiah bought the Churchyard Works, in the estate house of which he had been born, and installed his nephew, Thomas, there, leasing it on Thomas's death to yet another distant relative, Joseph Wedgwood, in 1788.

When you say Wedgwood in Burslem, you have to know which one you mean. The site of the Brick House—or Bell—works is now occupied by the Wedgwood Memorial Institute fronting on to Queen Street. What a building! Perhaps the most pleasing aspect of the endeavour is the care taken over the design. It was thrown open to public competition in 1860, and was originally won by a Mr Nichols of Wolverhampton. His elevation did not really do justice to the Potteries' greatest son, and after three years of trial and error he came up with a design including terracotta panels. (In the potteries, they call terracotta 'brick'.) The competition was reopened and this time the joint designs of Robert Edgar and John Lockwood Kipling were accepted. The result is a most exceptional building, generally described as 'Venetian Gothic'. The front has ten window-bays on the ground floor, five each side of a magnificent doorway topped off with a statue of Josiah, executed by Rowland Morris, no less. Spreading out on either side, and at the feet of the master, are ten panels showing the pottery industry as it was at the height of Victoria's reign. These splendid panels really do deserve a long, hard look. Designed by Matthew Eldon and made, again, by Rowland Morris, among others, they give a dazzling insight into the industry. Flanking the great man are twelve of the largest tiles ever. Representing the months of the year, they are the most sumptuous facing bricks imaginable. Above all this are mosaic panels of the signs of the zodiac. More in keeping with the Doge's Palace than a potters' institute, the mosaics have not worn as well as the rest, but the whole thing will reward a close scrutiny from the steps of the School of Art opposite, itself a jolly piece of Edwardian architectural adventure.

Sharing the ancient site with the Institute, which is now a library, is the Covered Market; again from the eighteen seventies it is another bit of Gothic with nicely pointed arches along the street. All the arches have suffered some indignity in the past century, but it is easy enough to detect the original design, which must have been very gay. Inside there is some fine Gothic-Revival cast-iron which makes the purchase of good quality vegetables at reasonable prices even more of a pleasure.

Actually, there is plenty of cast-iron in the Potteries area, and a lot of it is very beautiful. Most of what there used to be outside was pillaged during the Second World war, the same as anyone else's, but plenty survives as in the Burslem market and old town hall. It is helpful to remember the important part played by the long history of iron-working in the area when we consider the existence of the Pottery industry. The enormous growth of manufacturing could not have been achieved without the necessary machinery, however simple it might seem by modern standards, and it had to be produced locally since, before the building of the canal, importation of large quantities of iron was impossible. A very strong light engineering industry grew up alongside the pottery industry, in response to the same geological and economic stimuli.

Not that the Potteries is all factories and mines. In the same block as the Market and Institute, but on the north side, facing the old Town Hall, is the Leopard Hotel. It is early eighteenth century, its square, early-Georgian front having been given more room by the addition of two three-storey semi-circular bays on either side of the door in the 1830s. Inns, hotels and taverns of all kinds played an unusually important part in the commercial and industrial life of North Staffordshire in the early days. This arose from the system of employment and the labour-rich nature of the industry.

The owner was responsible for supplying the raw materials, choosing designs and decoration, marketing the finished 'ware' and paying a small number of skilled operatives. These operatives were all called 'potters', despite their widely varying skills. One such, Anne Griffiths, giving evidence to the Children's Employment Commission in 1842, said, "I have been a potter as painter seventeen years."

The ware was formed either by 'throwing' on the potter's wheel, or by being moulded on a shaped profile. Whatever process was used, and there were others later, the potters produced immense quantities of fragile pieces all needing to be transported with great care to warmed drying rooms to 'cure' before firing. Plates, varying in size from twenty-four inch meat plates to 'muffins' of less than seven inches, were carried, one at a time, complete with their plaster or fire-clay mould, into the heated sheds by small boys. Thrown-ware, like jugs, cups and bowls, was carried by the dozen on planks borne on the heads of older children or unskilled adults. Since the output of a single works was measured in hundreds of dozens daily—and still is—there were large numbers of these human conveyor belts at work.

The 'potter' was paid for the number of pieces produced and he in turn paid the 'runners' for the number of pieces that arrived safely at their destinations. The potter was paid in large coin and had to obtain 'change' before he could in turn pay out. This change was made at the nearby hostelry, whose incumbent was careful to see that a fair proportion of it went back into his own pockets. As recently as 1842, whence that report gives such valuable information, children aged eight to thirteen years got two shillings and a halfpenny (just over ten pence) for seventy-two hours work, and had to wait until the early hours of the morning after pay-day to get it. Shaw, in his *History of the Staffordshire Potteries*, published in 1829, informs us that seven-year-old boys in 1766 had between fourpence and sixpence a week. Shaw's fascinating book has to be read with extreme caution, as far as facts are concerned, but the flavour is there, and it makes for lively reading.

Apart from their financial activities the pubs and inns provided all the usual services for the work force. The people, who were called to work quarter of an hour before it was light enough to see, and went home when it was too dark to work, were given half an hour for breakfast at eight-thirty and an hour for 'dinner' at noon. Much of what they consumed was obtained from the nearby tavern, although some of the better-paid craftsmen and paintresses had food brought to them from their nearby homes by their children. There was always somewhere to heat it up on a pot-bank. Those not fortunate enough to have such service, and in an area renowned for teetotal nonconformity, were served by pie-shops and oat-cake bakeries. The oat-cake, which is not a cake, will need to be carefully explained to strangers to the area. It is nothing like the Scottish oat-cake, but is rather like a brown and knobbly pancake made from draught-porridge. Incredibly economical to produce, oat-cakes are very nourishing and sustaining. They are a symbol of the isolation and conservatism of the valley, since they appear to be an iron-age survival. When I, a 'furriner from off', first arrived, I encountered my first oat-cake, quite unaware of its true purpose and nature. Cold, it is tough and leathery. Since they were in a pottery warehouse in substantial numbers, they were obviously not wash-leathers, and anyway the windows had not been washed for at least two hundred years. Having heard, during a thorough if rather capricious education in Staffordshire history, of the 'poor crate-men', it seemed reasonable to suppose that they might be some kind of prefabricated separator for plates and so forth, which could, in some dire emergency, also be used as food to sustain these human pack animals on their journey to the distant markets.

In fact, eaten hot with bacon or fried cheese they have no equal, especially for filling the bottomless pit that teenage boys call a stomach.

Ryles' *Map of Burslem about the Year 1720* shows twenty-two alehouses—(and one exciseman)—out of 147 identified properties. One of the alehouses, run by a George Norbury, was the Packhorse in what is still called Packhorse Lane. This is supposed to be the route by which the 'poor crate-men' left the busy town for the markets of the world. One wonders how it came to be stopped-up and replaced by Newcastle Street, which seems to be a much more sensible route from the town. This ancient lane went right through the enormous pot-works of Enoch Wood, part of which still stands, on the left of Fountain Place. At its height, in the early 1790s, it covered most of the area enclosed by Westport Road, Hall Street, Blake Street and Newcastle Street. If contemporary engravings are to be believed it was surrounded by a high, crenellated wall, the battlements being continued on some of the buildings. Fountain Place works, as it was called, covered the sites of four other potteries and was built by Enoch Wood, son of the modeller Aaron Wood. This family is yet another example of the continuity of the industry. Enoch was one of a long line who are still at it in Burslem, and their ware has always been known for its variety and high quality, especially of form. The quantity and variety of production can be guessed from the number of different chimneys, thirty or thereabouts, shown in the pictures. To a limited extent the style of chimney depends on the sort of kiln it mothers, and that decides what is made in it. By the time Enoch built his place ware was being fired twice, once to turn it into ceramic and yet again to decorate and glaze it. The decoration might be moulded, painted or printed, or a combination of any or all of them.

It is not too profitable to try to date pottery processes, and broad generalisations, hinting that by such a date a process had been perfected, are quite common. For instance, the pots of the period between the 1660s and 1720s were supposed to be very ornate, largely to disguise the dark colour of the clay. Redevelopment allows excavation occasionally, and, in 1966 the destruction of the 1839 façade of Samuel Alcock's works along with part of the Royal Venton Pottery and some houses and shops, gave scope for investigation on a site occupied by William Burns Pot Works in the early eighteenth century. This is on the other side of the road from Enoch Wood's Fountain Place works, and a little higher up the hill. A large number of pots were found by the diggers, directed by Mr J. H. Kelly, the City Museum's archaeologist, all from around the same date, and many were wasters from the same firing. Wasters are pots that have gone wrong for various reasons and have to be destroyed. In this instance they were obviously buried near to where they were made.

Very few of them are over-ornate, and the bulk are decorated with slip in a great variety of restrained designs well suited to the shape and texture of the ware. Slip decoration is rather like an icing sugar technique in which liquid clay—'slip'—is trailed over the surface from a can equipped with a spout and a limited air supply. The result is very like a confectioner's art, and the rich brown of the body, with the creamy coloured slip, adds to the illusion. Forms of decoration can be seen on pots of this period, which continue for over a century. In fact, we can see in it the beginning of a Staffordshire style.

Another myth was that the slipware process was brought into North Staffordshire from Wrotham in Kent. Excavations in Woodbank Street, Burslem in 1975, showed that this sort of ware was being made there in the days before the Civil War, and whatever was happening in other parts of the kingdom, it was developed independently in the Potteries. It also means that some museum specimens may be at least fifty years older than was thought.

W. S. Landor wrote of Crabbe's poems that they were

> Truths written with a tuppeny nail,
> And scratched on a rough mud wall.

This is very unkind to the botanical clergyman, but it describes the decoration of the common pots rather well. We are accustomed to the endlessly refined, sophisticated product that has developed since the mechanisation of the industry robbed its end result of its soul. But, before everyone could read and write, there were some few who could paint and draw. Many of these were in the potteries, and many worked on the fine, effete china of that glorious period of Farmer George the third of our German kings and his ridiculous son, the Regent.

Most of the pieces that survive are the cherished possessions of the wealthy, carefully preserved as heirlooms. We should not be deceived into thinking that they are representative of the bulk of the pieces produced. Pots are everyday things in the main, made to be used, broken and replaced. Millions of pieces of lovely, lively, earthy peasant pots lie smashed and unrecognised in the top couple of feet of soil.

The earliest of the common pots were simply splashed with a bit of colour, and glaze, often accompanied by decoration cut into the clay; that was in the days of Robin Hood. Most of the early pieces were jugs and cooking pots. As the country grew up the variety of shapes and extent of decoration increased, until, by the time of the Tudor monarchs, drinking vessels, often covered in a lovely lustrous black glaze, were in use. As the number of these large cups increased, so did the number of handles they had. Usually with three and often more handles, they were called tygs. Presumably they were made in that way to make it easier to pass them round the table in that fine sense of community expressed by so many drinking songs.

As our Celtic monarchs gave way to continental ones, so

the range of types and decoration increased. Plates and dishes replaced the table-breads and pies of earlier days, and I suspect that lead poisoning replaced food poisoning as a major population controller, especially amongst the better-off. Still, with life expectancy being rather low, St Peter probably got them before the lead in their broth. The brief popularity of safe, salt-glazed ware certainly coincided with a period of population increase and expansion, just as it coincided with the introduction of tea and coffee drinking on a large scale. That gave an enormous impetus to the demand for the local product. Much of the excellent salt-glazed ware had a scratched decoration into which cobalt blue was rubbed, but it is a paradox that, having developed a white body to compete with the foreign opposition, the people of the potteries had become so used to brown pots that they frequently hid the nice white stoneware with iron pigments.

Having developed and standardised the material, which remains little altered to this day, the ceramic industry as a whole made very little progress in the means of producing and decorating the ware. Up to the 1750s, they were tied to hand processes of throwing on the wheel, pressing clay into moulds and joining the parts together, and painting on the surface in a very limited range of colours obtained from the oxides of common metals. Cobalt gave them blue, copper green, iron brown and sometimes red or black, manganese produced purply brown and cadmium a rather pleasant pale yellow. They could also mix these colours with varying success. All their pigments were soluble to some extent in the thick, lead glaze commonly used. These mixtures were absorbed through the skin of the operatives who 'dipped' the ware into a creamy brew of white lead, colour, ground flint and clay. They also breathed it in and ate it with their food, since the dust settled everywhere. This resulted in a hardening of the hands and arms until they were completely useless, together with a kind of sleeping-sickness. The whole obscene disease was known as 'potters' rot'. How generation after generation were persuaded to follow one another into the living hell of the dipper's shops it is difficult to imagine. Not that lead was the only enemy. The ceramic industry deals in a variety of fine powders, variously mixed into 'body'. All of the powders or 'dusts' are injurious to the lungs; some, like crushed flint, act like minute razor-blades, cutting and grinding the delicate tissues into a suffocating pink froth. The world little realises the pain, horror and stark degradation that were such an essential ingredient of the beautiful articles bought so cheaply and paid for at such a terrible cost. Everyone was rightly incensed when chlorine gas was used on soldiers in war. Who can count or who ever cared about the hundreds who suffered the same torments from the chlorine poured from the mouths of kilns when firing salt-glaze? In the Potteries, men and women are still choking to death on dust accumulated, little by little, in a useful, pitifully undervalued lifetime devoted to the creation of articles of great utility and considerable beauty.

Not that the employers were unaware of the suffering, and many would have liked to have done something about it, if only they had known how. Wedgwood's lovely Queen's Ware was especially dangerous both to produce and to use, and Josiah wrote, on 22nd August, 1773, "I will try in earnest to make a glaze without lead, and if I succeed I will certainly advertise it".

In fact, the answer had been found nearly a quarter of a century before, when Enoch Booth had mixed ground flint with lead glaze. The purpose of the lead is to reduce the melting-point of the glaze material, a process called 'fluxing'. and it was found, by a gradual process of experiment, that other materials would do the same job less dangerously. Amongst those tried with varying degrees of success were Cornish Stone, borax, whiting and felspar. Since most of these fluxing agents had to be bought from specialist suppliers they were expensive. This did not recommend their use to potters who were always cost-conscious.

Although flint was still being crushed or stamped in the dry state in the 1850s, Thomas Benson, of Newcastle-under-Lyme, had patented a process for grinding flints in water by 1726. To the lasting credit of the manufacturers, most of them had turned to the wet grinding within a few years. In effect, only the biggest firms ground their own materials; the smaller ones bought what they needed from their larger neighbours or from the growing number of specialist suppliers.

It was to accomplish this grinding that Enoch Wood built a windmill on his 'bank'. This useful device ground flints, mixed clay, ground colours and raised water. It was Enoch's mill that gave Fountain Place its name, indirectly. Like all of the great Potters, Enoch was public spirited to a degree, and his windmill—you can still find parts of the base at the blocked-up end of Packhorse Lane—raised more water than he could use. So he installed a fountain outside the entrance to his works and supplied water to the public for a small charge. This was a great benefaction. The two principal springs supplying Burslem were at the bottom of the ridge near the church and were contaminated by the contents of the graveyard. The church of John the Baptist is well worth a closer look. It has a tower on the west end, a nave and an apsidal chancel. The low, squat stone tower was built around 1536 as we know from a bequest of one John Tunstall, who gave money towards the "building of a steeple at Burslem church" in that year. Despite the romantic hopes of some parishioners, the Tudor-arched doorway in the west wall, surmounted by a three-light window, and the belfry stage with its three-light, late Perpendicular windows, support the documentary evidence. The chancel was originally a timber-framed building, in keeping with the domestic architecture of the period. That was burned down in 1717, complete with its thatch. It was then that the present brick nave was built. The population had grown so much by 1788 that the church had to be lengthened by the addition of the apsidal chancel, designed by Thomas Sherwin, and set in motion by the Churchwarden—Enoch Wood. It cost £700, so it was done in real style.

The chancel has a modern tablet erected to members of the Adams family from the fifteenth to the seventeenth centuries. Yet another Enoch Wood was responsible for making two fine terracotta panels in the chancel. His figure of Christ was done when he was fifteen years old. The other *Descent from the Cross* was modelled when he was eighteen.

In the cemetery many of the graves of famous potters are of interest, but pride of place must go to two rather special tombs. One is an ancient stone coffin from Hulton Abbey, tried on for size by all the small boys of the area, and a table tomb with its axis at ninety degrees to the others. This last is the tomb of Margaret Leigh, called 'Peggy Lee' by some. She lived in a house called Jackfield in the Hamil area of Sneyd. This was the home of the Leigh family, who had been there since the early 1600s. Her fame, or infamy, as the Burslem Witch had kept the estate isolated especially after dark. She was credited with all the usual powers, including a disposition to inflict a murrain on other people's flocks, and a nasty proclivity for keeping people's chimneys from drawing. Quite a handicap in an area rich in outsize chimneys. It is rather sweet that Burslem should have a witch in the 1730s, when everyone else had given them up fifty years before.

However, her crepuscular activities were not confined to her natural life. The family had to be pretty lavish with their entertainment in order to give the old girl a proper send-off, and since the parson, Mr Spencer, was an amiable drunk, the burial was quite a lively affair, and fortunately down-hill. They would have stumbled past what is now the park, along a bit of Moorland Road, past the Red Lion and Big House, into what is now Queen Street, down Bourne's Bank, over the stream to the churchyard. She was interred east-west in the family plot on the south side of the church, and off they went to drink her health. Again when the party arrived back at the Hamil they found the lady they had just buried sitting in the chimney corner knitting away, nineteen to the dozen. The party was dismayed. All that wasted effort. Parson Spencer was contacted and agreed to lay the 'bogut', provided he had the help of some clerical neighbours. They went off at midnight with bell, book and candle, sexton and clerk. When they arrived the other clergy ran off in fright, leaving the three locals to do the fell deed. They dug up the body, cut another grave at right angles to the first and popped her back in "in the shape of a blackbird", to lie for seven years. No one ever claims to have seen her again, although her presence still creates an occasional stir. Her mother outlived her by twenty years and was buried in the same grave, in the same alignment, presumably to hold her down. The date of the burial of the Burslem Witch was 1st April, 1748.

There were no witches buried by the other public spring, lower down the hill on Grange Farm, Cobridge. This was a very small spring and quite a long way from town, but Enoch Wood replaced his windmill with the wonder of the age, a steam engine, in 1798, and provided a public supply of water free of charge, rendering the springs obsolete. Although the use of water in mixing pot-clays is obvious, the 1750s brought another demand for water, as well as more burning and grinding.

In 1745, or thereabouts, a Burslem potter, Ralph Daniel, (these Daniels keep cropping up) 'Rafy Dennil' of Cobridge, went to France to see how they produced their figures. British potters were deeply mistrusted in France but by some subterfuge 'old Rafy' managed to get inside one of their factories and found that they used gypsum for their moulds. North Staffs Pottery Masters were deliriously happy about this piece of eighteenth-century espionage, and set to have models carved from 'plaster stone', which was readily available from the area of Burton-on-Trent, and not too far to carry it. Of course, carving the stone was no use and a further trip was needed to disclose the full secret: that the stone had to be calcined and ground like flint to make it into a powder to be mixed with water and cast into the desired moulds. This added yet another highly skilled operation to the already impressive list of ceramic trades, that of mould making, and led to a monumental increase in the output of the industry. Modellers, like Aaron Wood, became increasingly important, their original figures representing a high achievement in sculpture often overlooked in judging the result two hundred years or so afterwards. This episode brought the term 'plaster of Paris' to our language and opened the door to further improvements in potting.

Whatever route takes you into Burslem, there are any number of ways out. Visitors are recommended to try them all, since all have fascinating aspects that could never be revealed by spinning round on one's axis in front of a town hall or whatever.

If you have been there long enough to be leaving, you will have already discovered that the real name of the place is 'Boszlum'.

Northwards from 'Bozlum' takes you to Tunstall, 'Tunster', if you need to ask the way, and more of that later. That would take you along either the Scotia Road, or, preferably, through Fountain Place, past one of the most magnificent chapels imaginable. This is the Hill Top Methodist Church and Sunday School. It was built of brick with stone dressings in 1836–7, to the design of Samuel Sant. The first floor entrance has a double flight of steps leading through the central opening of seven bays, supported by Doric columns. The portico is topped off by a tall storey of three central round-headed windows and a classical pediment. Demolition of adjacent premises has improved the view recently and it is worth the effort to climb the steps for a look around the hill top.

In *Clayhanger*, Arnold Bennett says that he could see farms and cornfields, as well as the "forlorn oddness in that foul arena of manufacture", from the top of the hill; and also "the high spire of the Evangelical church, the low spire of the Church of genuflexions". Going down Hall Street, from Hill Top westwards, you will pass the "Church of genuflexions", the church and school of St Joseph, ministering to a large and healthy Roman Catholic congregation. What an intolerant man Arnold Bennett was. At the bottom of the hill, the "Evangelical church", St Paul's, is no more. Its lovely Hollington stone was no match for Potteries pollution and it had to come down. It has been replaced by what appears to be an outsize hot-dog stall. The substantial Victorian terraced houses that once formed a guard of honour on its southern side have been replaced by terraces rather similar to the church. On the bright side for those who mourn the passing of Victorian views, is that the new buildings should not suffer from mining subsidence, since, unlike their predecessors, they are equipped with a concrete raft. This is a device ensuring the uniform collapse of a building without those unsightly cracks in the masonry—where there is any—which used to disfigure old buildings in a coal-mining area. Now they will simply sink, all in one piece, like Wedgwood's Etruria factory, to vanish without trace.

Going east, or almost, from the Market Place along Moorland Road takes us over the old Knotty Railway line, now part of an internationally acclaimed reclamation scheme, past the park, through the village of Smallthorne and on to Ford Green. There is another railway line to cross and on the right is the site of the enormous Iron Works. On the left is one of the biggest surprises an ancient industrial town can give. Ford Green Hall. This is said to be the oldest house in the city, and is, not surprisingly, one of the few half-timbered houses in the Potteries. The hall is now a 'Folk Museum', in the care of the City Museums Department. It is very well cared for and maintained, vast sums of money and no little ingenuity having been expended on preserving it, not only from the ravages of time, but also from flooding. This flooding happened after mining "too low down, brought water from too high up" as it was explained locally.

Ford Green Hall really is an Elizabethan manor house, built about the time of the Spanish Armada, and it is easy to pick out the original central, half-timbered portion from the brick wings which were added in 1734 to replace the older out-buildings, destroyed, as usual, by fire. However anyone slept peacefully in such a combustible mixture as wood and straw is beyond my comprehension, especially as the early ones had fires built in the middle of the floor and no chimneys.

The house has been furnished with seventeenth- and eighteenth-century furniture, with the odd piece from earlier times for added interest. Outside there is a dove-cote to remind us of the no-nonsense providence of our forebears, and a special attraction is a scold's bridle, to remind us of their thoughtless, chauvinistic barbarism. Some of the magnificent items from the City Museum are displayed here from time to time, and visitors amuse themselves for hours deciphering the messages scratched on the tiny panes of the leaded lights.

Moving south from the market place in Burslem takes us down the Waterloo Road now. In the past our way would have been down Bourne's Bank and past the church, down Nile Street or past the 'Hell Hole', which was a court of squalid houses in the last century; their nature is evident from their name, some traces of the brick-work can still be seen, but the Potteries gave up the worst of its slums many years ago. That way leads us through the village of Cobridge and past their park.

Arnold Bennett called this place 'Bleak Ridge', but it never was bleak and it is not on a ridge. It used to be the Vill of Rushton Grange and was a marshy area where the Biddulphs once went wild-fowling. It was an early Saxon settlement near the Fowlea Brook, in the valley. The Grange was carved out on the slopes to the east by the Cistercian monks of Hulton Abbey some time before 1235. The Biddulph family were Roman Catholics and after the Civil War they leased the Grange to the Bagnalls, also Catholics, who made more of those butter pots there, and who heard Mass regularly during the seventeenth and eighteenth centuries. One advantage of the Catholic settlement of Rushton was that, during the plague which struck Burslem in 1647, the Biddulph's chaplain ministered to the victims with considerable success. This did not stop a Protestant mob sacking the Grange in 1688, in that fine fever of patriotism inspired by our change-over to a continental monarchy. The chapel at the Grange had become "a mere thatched shed" by 1820, by which time, despite some problems during the Gordon Riots a new church had successively grown to quite a size. The church was extended and improved many times up to the 1880s, but had to be pulled down in 1936 because of damage caused by mining subsidence. The present brick church and presbytery were built on the same site in 1937; that is St Peter's in Waterloo Road.

Started to commemorate the "famous victory" of 1815, and finished in 1817, Waterloo Road is a splendid mixture, and covers a lot of old ground. It starts at the George Hotel—the Dragon of Bennett's novels—a late 1920s building in the then-popular, neo-Georgian style much loved of G.P.O. architects, and replaced an earlier building of great charm. Moving down the road we are presented with a veritable parade of architectural history. At the northern end there are a number of small buildings from the end of the eighteenth century. The three-storied, bow-windowed American Hotel is rather like the old Leopard and was built about the same time. As we go further from Burslem the houses become more recent and much larger. This is the first middle-class area in Burslem, and contains a very good example of a mid-nineteenth-century terrace in Camoys Terrace. Baron Camoys was a descendant of the Biddulphs of the Grange, and his commemorative terrace is brick mock-tudor. What is now the Russell Hotel used to be a stuccoed private house in the Italian style.

On the left is number 205 Waterloo Road, one of the many homes of the Bennett family. Fans may like to know that he lived in Burslem from 1875 to 1878, first in Dane Street, then in Newport Lane, then at 198 Waterloo Road. Lastly number 205, which has been the Arnold Bennett Museum since 1960. However, he was born in Hanley.

II

HANLEY

NEXT door to Burslem, down the valley is Hanley. Here is the Potteries bright, modern shopping centre, law courts, the brand new, sky-scraper administrative block known officially as Unity House, and widely, if unofficially, as Hanley Castle. To get the best out of that you have to know how to say it. The two 'As' are the same as in 'cattle'.

Hanley is made up of the two moorland hamlets of Upper Green and Lower Green, and the old Vill of Hanley. The name is early English and was spelled 'Hanlih' in 1227. The name means either a high wood or a clearing in that wood. The upper Green was the area around Town Road and the lower Green was around the Market Square. Together they were known as Hanley Green. The town now consists of that lot and also the old residential area of Shelton, where a splendid Victorian–Gothic tower tops an outstanding church in an area where outstanding churches and towers are almost commonplace. In 1905 Birches Head and part of Sneyd Green were added to Hanley.

Shelton Church overlooks the valley in which Wedgwood built his Etruria and his fine Georgian house, now the offices of the Shelton Bar Iron and Steel Works, whose furnaces paint the night sky of the Towns a lurid orange. A tantalising beacon to those men and families whose fight to retain their livelihood has already become an industrial folk-legend. "Shelton Bar", by the way, is not the crenellated gate of an ancient town fortress, like Perry Bar; it comes from the product, bar-iron, to distinguish it from a sheet-rolling mill.

Born in 1683, in Shelton Old Hall, which used to stand on the high ground south of St Mark's Church, was Elijah

Fenton, who may not have been England's greatest poet, but may well have been her laziest man. He went to the Grammar School in the nearby metropolis of Newcastle-under-Lyme, and was the friend of another of Staffordshire's literary giants, Dr Johnson, of Lichfield. He translated four books of the *Odyssey* for Pope, who befriended him, and generally made himself agreeable in the best houses. How he maintained the stamina to write his interesting and perceptive *Life of Milton* is a mystery. Dr Johnson said of him that he liked to lie in bed and be fed from a spoon. Elijah's epitaph, written by Pope, says that he was:

> Foe to loud praise, and friend to learned ease,
> Content with science in the vale of peace.

That must have been what Etruria seemed to be when Josiah Wedgwood bought the Ridge House Estate in Shelton Township in 1767. There were two or three large houses in Shelton and a few small houses and cottages on the ridge, but very little happened there until Josiah took an interest. He had outgrown his old works in Burslem, and was going to build the world's most modern pottery factory on his newly acquired estate. So that his workers did not have to trail down the hills from their villages and arrive tired out at the works, he was going to build them homes right there, on the door-step, so that they would arrive bright-eyed at the crack of dawn.

The *Gentleman's Magazine* of 1794 called it "a colony newly raised where clay-built man subsists on clay".

It was not that simple. In the middle of the eighteenth century, French pottery was a good deal better than English, and large quantities of it were imported for the tables and sideboards of the wealthy. A number of Staffordshire claybakers, Wedgwood amongst them, experimented with great urgency to overcome their commercial disadvantage. The delicate cream-coloured ware, light as cardboard, that gave Mr Wedgwood the distinction of being Her Majesty's Potter, was his passport to affluence as well as influence. He first caught Queen Charlotte's attention by presenting her with a 'candle-service'—I suppose that means a dressing-table set—when she was in the process of presenting the nation with the infant who was to become George IV. It was not long before the flow of crockery was reversed and we were sending it to the continent, even as far as Russia.

At about this time the problems of the transportation of his fragile and precious ware became most urgent. Perhaps it was because he was one of thirteen children, and the youngest of them, that he seemed to think in terms of concerted action and corporate responsibility from the start. Throughout his life he identified his prosperity with that of his fellow potters. Whilst they regarded each other as competitors and enemies, Josiah saw that his part of the industry could only survive if the industry as a whole was healthy, and many of his efforts did not benefit him directly at all. One that did was his championing of the turnpiking of the old lanes. The best way out was still on foot, human or otherwise, and this severely limited the loads. The carts either bounced over boulders and ruts in the dry, breaking the crates and their contents, or they were mired-in in the wet and created huge bottlenecks in communication. The same was true for the essential raw materials they all needed to import.

There were good roads nearby at Newcastle-under-Lyme, but even the four and a half miles to reach these was tiresome and expensive to negotiate. What the potters wanted was to reach navigable water by the quickest route, and with cart-loads, not bags-full. This water was on the Weaver at Winsford and thence to the Mersey and the overseas markets. The other way was to Willington Ferry on the Trent for destinations south and east.

The potters, under Josiah's captaincy, petitioned Parliament to build turnpike roads. The corporation of Newcastle-under-Lyme opposed this move since it meant a loss of revenue to the town—they had their own toll-gates to look out for—and also promise of growth in neighbours which might, and of course eventually did, usurp much of their influence.

A House of Commons Committee called for evidence from the petitioners in 1763, and Wedgwood told them that the roads were "in very bad condition, narrow in some parts, and in the Winter Season impassable in many Places".

Parliament granted their Bill and turnpikes were constructed giving them access to both of the water-ways they desired to use.

The new roads did not solve the problem; they merely eased it. The turnpikes from Burslem through Church Lawton to the Weaver, and through Newcastle-under-Lyme to Uttoxeter and beyond were only a stop-gap, and of more use to local than to national traffic. A scheme to join the Mersey and the Trent by a canal was soon conceived. This appealed to Josiah, and he was its most enthusiastic supporter.

The first meeting of those interested in the proposed canal was held at the Leopard in Burslem, in March, 1765. Earl Gower, noting Josiah's enthusiasm, asked what sum he was prepared to put up. Jos answered that he would subscribe £1,000, and take a good proportion of the shares as well. This engendered spontaneous excitement in the crowd, all rushing to donate money and lend a hand. Wedgwood must have wondered if he was going to have to dig it himself when he was, at the inauguration of the work, given a spade and invited to begin. However, he was simply being honoured in the cutting of the first sod.

It was when he realised where the canal would be going that Josiah decided to invest in land at Shelton through which the canal must pass. The canal was to be called the Grand Trunk Canal, because it was envisaged that branches would follow the contour in every direction, spreading pottery far and wide. In the end Trent and Mersey Canal proved both adequate and accurate. It runs from Preston Brook on the older Bridgwater Canal to Derwentmouth on the Trent. From there the river is navigable. The work took from 1769 to 1777 to complete, although it stretched from Stone to Derwentmouth by 1771. The engineer responsible was James Brindley, who, sadly did not live to see it finished. He had other irons in the fire and, by 1790 a system of canals existed linking the deeper parts of the Trent, the Mersey, the Severn and the Thames. Throughout the Middle Ages the old roads had met and crossed in Staffordshire, always avoiding the Six Towns. Now the new communications network met, once again, in Staffordshire; one such junction was, and still is, at Great Haywood, near Stafford. And this time the route lay through the Fowlea Valley. The Potteries was wide open; the sky was the limit.

Josiah opened his new pot-works in 1769. On the high ground above the canal he built Etruria Hall, on the north side of the Newcastle road. The house overlooked the new works. There is no truth in the story that the works was connected to the house by an underground passage, though it did have fine, vaulted cellars in which Josiah built his laboratory and personal workshops. Above these cellars rose a standard sort of Georgian, three-storied, brick-built, stone-dressed, house. The three centre bays of five projected, Parthenon-like, and Parthenon-like were surmounted with an Attic pediment. In fact they still are. It is not too hard to pick out the original house from amongst the additions, which, in the main amount to two low wings connecting two-storey blocks of unimaginable purpose to the original building. Because of the hideous depredations of later and more recent tenants, it is most difficult to picture the lovely grounds as they were in Etruria's infancy. Capability Brown did not lay out the park; Capable Josiah Wedgwood did. The potter and the potterer were acquainted and no doubt discussed gardens in general, but one hardly sees Mr Wedgwood, the Queen's Patentee, allowing anyone else to decide the disposition of his new kingdom.

They all moved in ('flitted' in Burslem) in 1771, and just how lovely it all was can be seen in George Stubbs' paintings of them. This was about the time that another art lover, Sir William Hamilton, British envoy in Naples, was making known the beauties of the so-called Etruscan ware. A number of pieces were imported, but, most of his collection was lost by shipwreck. The catalogues survived, however, and gave valuable inspiration to art in general and to potters in particular. They were in the process of refining their wares, and were short of something to copy. Natural, English style, genre decoration would not serve the rich. They asked for something more fashionable. A gimmick was required, and the Etruscan Vases provided it. Form, body, decoration, the lot. Wedgwood was not the only one who wanted to copy this superb ceramic material, but he was the most successful, commercially at any rate. His curiosity and scientific ability had led to many experiments for their own sake, which now paid dividends in pointing the way to new bodies for old pots.

Whether to encourage people to think they had really ancient pieces, or merely out of respect for the area in which his inspiration was found, Josiah named his estate Etruria, and so it remains to this day.

Wedgwood had known Thomas Bentley for some time, and had expended a great deal of effort in attempting to get this cultured and influential Liverpudlian to join him. In the spring of 1767, Josiah delivered an ultimatum to Bentley, who had been hedging for ages. He informed the reluctant merchant that he was overwhelmed with ideas for his works, which, he insisted had to be as pleasing to the mind as they were profitable to the purse. That was why Bentley was needed. He was a man of great ingenuity, was widely read, and had a large circle of friends amongst the wealthy and influential people of the court and of commerce. The one indispensable ingredient he could bring to the partnership was 'taste'. He could provide an air of respectability to what was looked upon as a rather distasteful, menial business. Wedgwood's salvo ended with a typical pot shot: "Either resolve quickly to join me yourself, or find me out another kindred spirit." Bentley joined. The partnership flourished. Thick and fast the novelties flooded from Etruria, down the canal to the ready hands of the whole civilised world. They even went to America. Wedgwood and Bentley was just beginning to blossom there when the American War of Independence broke out. There was a great deal of sympathy for the colonists in the Potteries, and much crockery was produced with pro-American slogans and with scenes pandering to American patriotism. Since North America had always been a lucrative market for our products, the potters were faced with quite a dilemma, and with Europe going up in French flames immediately after the War of Independence, any manufacturer who was not making arms faced a bleak future. The trick was to either capture or to flatter fashionable minds. The products of Etruria did this to a remarkable degree.

Not that all the Potteries products were so genteel, Nor were all the Wedgwoods so discriminating. In *An Asylum of Fugitive Pieces* published by Almon in the 1760s we read:

> I am told that a scoundrel of a potter, one Mr Wedgwood, is making 10,000 spitting-pots, and other Vile Utensils, with a figure of Mr Pitt in the bottom. Round the head is to be a motto—
>
> We will spit
> On Mr Pitt.
>
> And other such d—d rhymes suited to the uses of the different vessels.

The mind boggles. It is hard to escape the feeling that politics has lost something over the years.

Although Josiah may have been making spitting pots and other vile utensils, in competition with his relatives, he is better known for:

> Busts of Antiquity, urns, vases, mythological, classical, and original compositions in imitation of cameos, intaglios of antique gems, or heads for seals and rings, full sized heads of celebrated men, dead or living, or medallion likenesses, including complete series of sovereigns of England and France, and of the Popes, from Linus to Clement XIII (numbering 253, in progress)—all these interesting designs were executed for the purpose of attracting the attention of the wealthy and the curious in every part of Europe, where they were eagerly sought after, and obtained handsome prices.

Bentley, on Wedgwood's behalf, established a London showroom in Greek Street, Soho. This soon became a fashionable lounging-place for the rich young fops and boosted sales enormously. They vied with one another to have the latest piece from Etruria. Catalogues of their huge stock were printed in both English and French, giving an historical account of the originals as well as a glowing description of the copy offered for sale.

The most famous of all the products of this fabulous factory

are the copies of the Barberini or Portland Vase, long considered North Staffordshire's *chef-d'oeuvre*, and certainly the
most remarkable old crock. Josiah had great difficulties in
perfecting the work and commensurate pleasure in overcom-
ing them. The original vase was found in about 1663 in the
Monte del Grano, near Rome. It was filled with ashes and
enclosed in a marble sarcophagus decorated with classical
figures in high relief. The sarcophagus was put into the
Capitoline Gallery and the Barberini family claimed the vase
for their sideboard. Urban VIII was Pope at that time and was
one of the Barberinis. They regarded it as their most valuable
jewel amongst many, as well they might, and kept it for about
150 years. It then fell into the hands of Lady Hamilton's
husband, who gave it up to the Duchess of Portland. In 1786,
her "rich cabinet of curiosities" was sold, and the Duke of
Portland bought the Barberini Vase for one thousand guineas.
The noble gentleman lent it to Josiah who kept it for a year,
during which time he made a copy. He subsequently made
fifty more, each of which he sold for fifty guineas. A number
of editions have followed, none matching the first batch which
had "the testimony of the Duke of Portland, Sir Joseph Banks,
President of the Royal Society, the Earl of Leicester, President
of the Antiquarian Society, and Sir Joshua Reynolds, President
of the Royal Academy, to the exactitude of resemblance
between the original vase and his copy". What a line-up of
witnesses. Josiah does himself and his workpeople less than
justice by emphasising the resemblance between original and
copy. The Barberini Vase was broken by some lunatic and was
indifferently repaired. Recently some extra bits were discovered in a drawer at the British Museum, and it is now
impossible to tell it was ever even cracked. It remains a
sublime example of classical cameo glass.

Josiah's Portland Vase is the ultimate in ceramics. It is no
longer a copy; it is a pot portrait of the glass model, and like
any other portrait, exists as a work of art in its own right, to
stand or fall by its own merits. It is the later issues that are
copies; and beautiful though they are, they have the stiff,
awkward attitude of stand-ins and lack the vitality and fluid
sensuality of the first fifty. They remain a satisfactory sub-

stitute, and I hasten to confess that I should be quite satisfied with the substitute.

The firm used the mark Wedgwood and Bentley until the latter's untimely death in 1780. Josiah was very upset by the loss of his friend who had done so much for the ceramic industry. His contribution to the success of the Etruria venture was prodigious. He managed the London showrooms and the factory in Chelsea where the ware was 'enamelled'. The decorating process was very expensive and it made sense to carry the pots to the London centre of distribution 'in the white'—that is, undecorated—and to do the painting and so on there. In this way, any breakages were of the cheaper half of the article. What is not often realised, in stressing Bentley's undoubted social assets, is the fact that he had an unparalleled knowledge of machinery and raw materials, and was as useful in the workshop as in the salon. Not only this, he was a letter writer and pamphleteer of great influence, always ready to support his partner's projects; and his knowledge of Parliamentary procedures helped with both the land schemes and the protracted struggle to break monopolies of clay supplies, especially against Champion.

This last mentioned effort illustrates the public-spirited generosity these two men shared. Despite all the progress and improvements, the principal competition for the Potteries product was still continental and oriental porcelain or 'china'. Potters, local and otherwise, had experimented throughout the eighteenth century to find a means of producing it in quantity. Factories at Stratford-le-Bow and Chelsea, in London, had both had early success, with Chelsea well in the lead by 1750. The continental competition was very hard to beat since they enjoyed Royal patronage on a lavish scale. The enormous cost of Böttger's experiments, which were designed to discover how to make gold, and ended in the discovery of a porcelain body for Augustus the Strong, Elector of Saxony, earned the resulting pots the title 'The Bleeding-Bowl of Saxony'. They were truly magnificent and bankrupted the state. His stuff was so good it could be cut and polished like glass or stone. The first examples were red and he called them 'Jasper'. Elsewhere on the continent the search for the secrets of the manufacturing of the prestiginous porcelain was prosecuted with great intensity. The Italian factory at Doccia, near Florence, was supported by the fortune of the Ginori family. Opened in 1735, it had houses for the workpeople, a hospital and a school. The products were exquisite in decoration, though rather muddy in the body. The famous Capodimonte factory was paid for by Charles IV, King of Sicily, in 1743. In France, the Duc de Villeroy supported the factory at Mennecy which made some charming naturalistic figures and a large number of pretty trinkets for the furniture of ladies' boudoirs. This is hardly surprising since ladies' boudoirs were all the rage in eighteenth-century France. At Chantilly the Dubois family had a factory which was constantly inventing lovely decorations for their incomparable pots and they were responsible for 'seducing away' numbers of English workpeople. The King of France had them removed to the Royal Castle at Vincennes where they set up a porcelain factory, were sacked for their trouble, and replaced by one of their own workmen at the head of a company mysteriously called 'Charles Adam'. Madame Pompadour, the French king's mistress, was quite a champion of the arts, and had the Vincennes factory moved to the Village of Sèvres to be near her Château de Bellevue. They began 'making' in 1756, and are still hard at it.

Whilst the continental manufacturers all called their products 'porcelain', some were more like true porcelain from China than others. There is a distinction between true porcelain, which is called 'hard paste', and a substitute for it called 'soft paste'.

We now know that china was made from Kaolin, or white china clay, and petuntse or china stone, a type of felspar. But we did not know in England in the 1760s and the potters of Meissen kept it to themselves. Consequently a number of different experiments were made to reproduce the material. No analytical techniques were available to help so it was a case of trial and error—mostly error—and scores of people were ruined in the attempt. Since it was a translucent, white body that was needed, and clay is white and glass is transparent, most of the trials consisted of mixtures of glass and clay. The glass had to be ground up first and the process of first firing materials and then grinding them up for use is called 'fritting' and what you get is called 'frit'. It worked, after a fashion, but the pieces that either boiled or collapsed in the kiln far exceeded in number those which survived, and then they had to be decorated and heated-up again, causing even more losses. Little wonder that the hard-headed Staffordshire potters had no use for 'Chiney' ware. Two early attempts, one at Newcastle-under-Lyme from 1744–54, and the other at Longton Hall from 1749 to 1760, were not commercially successful, but are amongst the earliest of English attempts at porcelain, albeit with a soft, glassy body, rather like the better-known pieces of Bow.

Whilst all this was going on locally, others, elsewhere, were experimenting. William Cookworthy, a wholesale chemist and druggist of Plymouth, applied his good education, great intelligence and keen powers of observation to connect the china clay and china stone he had found in Tregonning Hill with what he had heard of Meissen. On 17th March 1768, he was granted a patent to make porcelain from Cornish stone and Cornish clay. His factory was at Plymouth. In the autumn of 1773 Richard Champion of Bristol, who had been experimenting with porcelain making, bought Cookworthy's patent and other rights, and proceeded to make the ware at his Bristol factory. This was a true hard-paste porcelain, although imperfectly glazed. In February 1775, Champion presented a petition to Parliament for an extension of Cookworthy's patent for a further term of fourteen years beyond the original fourteen. In this he encountered serious difficulties, the chief amongst them being the strong opposition of the Staffordshire earthenware makers, headed by Josiah Wedgwood. Josiah had no interest in making china himself, and his works never did so during his life time. He knew he could do best that which best he knew how to do. Their reason for resisting the Cookworthy patent was that, if it succeeded, they would be prohibited from using clay from Cornwall for their pots, although they had been doing so for nearly one hundred years. Wedgwood had tried out clay from as far away as America and China, and was not prepared to be deprived of supplies from nearer home. Champion's Bill got its three

Burslem Sunday School, also known as Hill Top Methodist Church

The Roman Catholic Church of St. Joseph, Arnold Bennett's "Church of Genuflexions"

Burslem's Old Town Hall, to the left is the new Town Hall, now the
'Queen's Theatre', and to the right is the Wedgwood's Big House,
now the Midland Bank

The tomb of Margaret Leigh who was buried on April Fool's Day, 1748, at right angles to everyone else

The Parish Church of St. John the Baptist whose Tudor tower is barely able to rise above the nearby 'bottle ovens'

Wood's factory at the top of Packhorse Lane as it was in the early
nineteenth century

The remnants of the building today

(*left*) "Clayhanger Country" – a bit of old Burslem where the Salvation Army and the pub stand side by side beneath a bottle oven. (*right*) Lyndhurst Street, terraces of bungalows are replacing terraces of houses

The Wedgwoods' Big House, unchanged apart from the signs since the middle of the eighteenth century

Middleport Pottery, a canal-side works which shares the rich soil of
the Fowlea Valley with some very neat allotments

Ford Green Hall – a pleasant surprise in an increasingly derelict
area

The impressive canal-side bottle-oven of the Price and Kensington
Potteries

The path of the old Loop Line makes a fascinating walk-way with a
wide variety of bridges and scenery

Cobridge Park

readings in the Commons, so Josiah and Bentley took it to the Lords. There it was amended to make it impossible for anyone to be prevented from using the Cornish materials, although they upheld Champion's right to control the recipe he had bought from Cookworthy. The Staffordshire potters agreed that it was fair for him to reserve to himself those rights into which such a lot of hard work had gone, and which had cost so much. Having lost his chance of a monopoly, Champion also lost heart and closed his works in 1778, only three years after his hollow victory in the House of Commons. He also recognised Wedgwood as the Prince of Potters, since it was to him he turned when he wanted to dispose of his secret of china making. True to form, Josiah drummed up a consortium of his closest friends and on his personal initiative effected the introductions which led to the purchasing of the patent and the formation of the New Hall Company in 1781.

Wedgwood's philosophy of life, at least as far as his equals was concerned, is summed up in a letter he wrote in 1792 to his nephew Thomas Byerley, who succeeded Josiah as Master of the Etruria works: "I shall let my actions speak for me and continue to perform what appears to be my duty to all around me. I shall certainly love my friends, and if I cannot arrive at the perfection of loving my enemies, I will not injure them, unless in my own defence."

That last sentence may explain the happenings of March 1783. The American War of Independence had just finished and trade was stagnant. France was bankrupted by her support of the American colonists; that venture is estimated to have cost her four hundred million dollars, and is one of the reasons why the American Revolution is sometimes described as having been lost by the British because "their Frenchmen beat our Germans". Anyway, France was on the verge of collapse and, whilst all Europe held their breath, and our cousins in the New World prepared to fend for themselves, our ports were idle and our people starving. Since the political organisation of labourers was still in its infancy, their only recourse to help was in rioting. The destitute of the Potteries were in step with the fashion of the day and were rioting away noisily, even at Etruria, where an enlightened master had built a "capital manufactory, and a continuous street of about 120 workmen's dwellings adjacent, with an inn and some houses of a better class, for farmers, clerks, and others". This was near the new canal, along which a boat had come, laden with flour and cheese. The barge stopped at the Etruria wharf, to unload its cargo for the Potteries, but the owners decided they could get a better price if the cargo went on to Manchester. As soon as the grocers of Hanley heard of this they told their customers who formed a posse to recapture the boat which had got as far as Longport. There they caught up with it, several hundred men, women and children. One of the men jumped into the boat, whereupon the bargee slashed the mooring lines and attacked the solitary member of the boarding party with his knife. The man avoided the thrust and many of the mob rushed to his rescue, with every intention of pitching the bargee into the Cut. They were restrained by an unnamed gentleman, and brought the barge back to Etruria in triumph. They pillaged the contents and stored them in one of Josiah's sheds. He was not at home at the time, but his eldest, Josiah II, his wife and son John all went down to harangue the rioters who left a few of their number guarding the goodies and went away quietly to their homes. The sentries went up to the Hall to ask for refreshments and seem to have been sympathetically received, even though they were scolded by Mrs Wedgwood. That was on 7th March. On 8th March the cargo was sold to the mob at bargain prices and the proceeds handed over to the boatman. Carried away with their success, they disposed of the cargo of yet another provisions boat in the same way. As an encore they tried to set fire to some of the houses, and the works seemed to be threatened. Josiah was home by now, and sent to Newcastle for help. Help arrived in the form of a company of the Welsh Fusiliers and a detachment of the Staffordshire Militia, under Major Sneyd. They charged the mob who prudently dispersed since what they wanted was food, not war, and two men, Stephen Barlow and Joseph Boulton, identified as ringleaders, were arrested and hanged.

What is often overlooked by romantic apologists for this period in our history is that Wedgwood was typical of the rising group of successful capitalists of the eighteenth century. He told Bentley that, in order to get his people to do what he wanted them to, he had not spared them in threats and would have thrashed them right heartily, if he could. He was an outstanding personality with an extraordinary gift for organisation, and was a great artist. But he was no different, in his attitude to his employees, from any of his contemporary industrial barons. In fact, his enormous business success, and the social power this brought him, made him even more of a tyrant towards his workers than many of his less wealthy colleagues. He had that sense of ownership of his employees that was a relic of the paternalism of the craft days of the industry, coupled with the power of a leading magnate of the new industrial aristocracy. An unbeatable combination not enjoyed by many of his fellows.

The application of his social influence resulted in an act of magnanimity towards his equals that was in sharp contrast to his treatment of those he called mere 'pottery hands'. That was in the affair of the Champion Patent, and the formation of the New Hall company. The confederates who formed the company were Sam Hollins, who made lovely red-ware teapots at Shelton and whose name is worth remembering; Charles Bagnall, also of Shelton, a practical potter and a member of an old local family; John Turner, who was every bit as good a potter as Josiah Wedgwood and whose works was at Lane End; Jacob Warburton, a master-potter of Hot Lane, Burslem, who lived to the ripe old age of eighty-six, despite all the odds against it; William Clowes of Port Hill, Newcastle, and who is shrouded in mystery; and Anthony Keeling of Tunstall, at whose works the first operations of the new company were conducted.

As was inevitable in a group formed from members of a class in which only self-seeking egoists survived, they fell out and Turner and Keeling sulked off to plough their own individual furrows. Since Mr Keeling and the new company were not on speaking terms, they had to remove their 'tranklements' from his premises and they took themselves off to Shelton Hall, afterwards called the New Hall. The Hall, which stood on the piece of ground between Hope Street,

Sampson Street, Century Street and New Hall Street, behind the lovely old Hope Street Church which is being pulled down to make way for yet another supermarket as I write this (1976), was being used as a pot-works by Thomas Palmer. Trading as Hollins, Warburton and Co. they made large quantities of true porcelain, decorated in a rather naïve way, first with sprays of really English flowers and later with print and enamel in the Chinese manner. No ware was more typical of the Staffordshire Potteries than New Hall, and very few bodies have taken the bright colours so well. One of their most successful ventures was the making and selling of glaze according to Champion's recipe. They went on with their hard-paste until 1812 when they went over to bone china. The works was constantly in production of one sort or another, under various owners until 1957.

Most of the big old pot-works that grew up in Hanley at the end of the eighteenth and early nineteenth centuries have been demolished. Architecturally they are not to be mourned since there are still many like them in the Six Towns, but one wonders how long anything will remain in the mania for rebuilding that even national economic chaos does not seem to curb. Of course, most of the important buildings in the Towns are replacements for earlier ones, and it may be that those now in the course of erection will inspire future generations to paroxysms of conservation fervour, but somehow I doubt it. On the other hand, contemporaries find it hard to consider architects as far-sighted visionaries. We can only see that a little bit of our consciousness has been chipped away and replaced by something we do not understand, and which is bound to be incongruous amongst all the older buildings with which we are so comfortable.

Mr J. W. Plant, who as City Architect and Reconstruction Officer, was responsible for so many of the changes, said that Hanley Town Centre was an "archipelago of island sites". This is very true, and it makes of Hanley a bewildering place. Redevelopment has emphasised the 'island' feeling by sticking to the old ones and building new ones. One of these new islands reinforces the sneaking suspicion that the architect might, after all, be on to something. This is the area that used to be Clementson Street and the top of Cannon Street, alongside the new museum and Warner Street. A new Library and Police Station have been built on the new island, and where the quaintly named Crown, Orb and Sceptre Streets used to stand in Unity House (Hanley Castle). Standing beside the Spitfire in its plastic greenhouse, with your back to what used to be John Street, looking between the buildings it looks all wrong. But, just by moving down between the buildings to a kind of a court-yard, looking towards Unity House an immediate impression of 'space 2,000' is created. Just for a split second you stand in the future, lost in an architect's grasp on a future reality. It is quite uncanny to be forced to realise that, despite the current incongruity, the planners are likely to be right.

After this excursion into the future it is quite a relief to find the Bethesda Church and Sunday School intact and thriving. Inside, the chapel is much as it was built in 1819, but the front, on Albion Street, was changed in 1859, by the addition of an impressive colonnade of Corinthian columns. They also put in an enlarged window over the colonnade and topped that off with a restrained cornice. The other windows around the church were put in in 1887. The Sunday School, which looks splendid across the graveyard, was also built in 1819, though it was lower then, the upper storey having been added in 1836. With its central pediment and beautifully proportioned roof, the school is as pleasant a piece of Georgian England as you will find anywhere. The cobbled entrance and stately tombs all conspire to make a memorable impression.

At the top of Albion Street is the Town Hall in Albion Square. It is not like any other town hall in the Potteries, which is hardly surprising, since it was built as a pub. It was opened in 1869 as the Queen's Hotel, and is more like a French *château* than an administrative building. Built of brick, which is eminently sensible in an area that makes such good bricks, it has the inevitable stone dressings, and comes out to meet the public on three broad bays whilst the dormer windows in its steep roof give it an air of watchful surprise. Tacked on to the back of the town hall is the Victoria Hall. Acoustically one of the best halls in the country, it was erected in 1887 to the designs of the borough surveyor, Mr Joseph Lobley. Opened in 1888 it has been the centre of a flourishing musical scene ever since. Elgar was content to come to conduct here, and gave the first performance of his *King Olaf* at the Victoria Hall in 1896; a joint tribute to the acoustics, the high standing of the choir, and the growing importance of the Potteries. It seems, somehow, very apt that Longfellow's words:

> The guests were loud, the ale was strong,
> King Olaf feasted late and long

should have been bellowed by the throats of good Potteries folk, long famed for their nonconformist temperance virtues in an auditorium leaned up against a town hall only recently converted from an hotel.

Perhaps less convincing, but just as great an accolade, was the first performance of Delius's *Sea Drift*, under Thomas Beecham in 1908. Delius had himself conducted a concert here earlier that year, and, discussing this and Delius's later torments with an aged member of that choir, I was vigorously informed that: "He didn't catch anything in 'Anlee!"

It is no surprise to find the Albion Hotel supervising an extraordinary, complex road junction in Albion Square; the hotel has long been the resort of concert-goers from the Victoria Hall, and it must be a great comfort to all concerned that the police station that stood between them and their refreshments has now been destroyed.

Diagonally opposite the Albion is Tontine Street, leading to Tontine Square. In Tontine Street is the Post Office, opened in October 1906 at a gala occasion by the Postmaster General, attended by a large crowd of important people, with an even larger crowd of presumably unimportant people looking on. A long procession headed by mounted police and consisting of civic dignitaries complete with Mayor; church dignitaries complete with Canons, the Royal Artillery Band, without canons, and most of the more successful shopkeepers—it was Thursday the 18th October—wound their way through the streets of Hanley to see the job done right. The august personage made his first speech in the main hall, calling upon the public to write more clearly because too many packages were going astray. He claimed that twenty-seven million were not delivered in the previous year. He made another speech at the Grand Hotel later, during a celebration luncheon, and the Post Office was considered well and truly opened.

The building was erected on land purchased by the post office in 1900 from Hanley Corporation who had owned it since 1876. Before the site was cleared it had for many years been the venue for Batty's Circus. How nice that the opening of their successor should have been attended by a circus just as 'batty'. It is a good building in the classical style, constructed from Hollington Stone, and the front has Ionic columns and pilasters supporting a triangular pediment. The theme is repeated in the two first floor windows of the slightly projecting end bays; these are topped off with a semicircular pediment in rather poor taste. However, the overall impression created is good, and the building is an ornament to the town.

One of the best things about the post office is to be seen when standing with your back to it. Opposite are some of the original low, brick built houses of the early nineteenth century. Although they are now converted into shops, their cottage-like charm is inextinguishable. Crossing Percy Street brings us to the old covered market, built in 1831 as a shambles—which originally meant a slaughter house—it is a grand old classical-type building and the front, facing Tontine Street, has a central, arched entrance surmounted by a stone turret and cupola. Flying away on each side of the front door are Doric colonnades with pedimented side wings. The heavy stone basement blocks are worn hollow by a combination of the sharpening of butchers' knives and the shuffling of six generations of little bottoms waiting for their mums to emerge from the shambles within.

Although few of the buildings of the first half of the nineteenth century remain, the 'islands' created by the ground-plan of the 1830s remains hardly altered. In *White's Directory* of 1834 for Staffordshire, Hanley was described as a large modern town, the largest in the Potteries and second in Staffordshire only to Wolverhampton, its streets were spacious and well-paved, its houses were neat and some of them were, like the public edifices, elegant. Small wonder then that this capital town of the whole Fowlea Valley should name a little group of its new streets after those of London. So we have Cheapside, Piccadilly, and Pall Mall, in a naïve apeing of a foppish, distant area by people who have rarely understood just how proud they ought to be of themselves.

I would have wanted them named Astbury Street, Whieldon Street and Wood Street, or something. But then, I am not a native, and like so many 'converts' am much more fanatical than those who chanced to be born here. The natives rarely know what treasures and privileges they have until it is too late and it is too late for some of them. This brings me to the late Mechanics Institution in Pall Mall. It is all too easy to confuse Institutions with the buildings built to house them. The original Mechanics Institution was founded in 1826 for "the promotion of useful knowledge among the working classes" and as such was a forerunner of the Workers' Educational Association, which has always been so strong in North Staffordshire. The originators of the 1826 venture were the Rev. Benjamin Vale, at that time curate of Stoke and later Rector of Longton, and Josiah Wedgwood II. They had support from two successive Dukes of Sutherland and a large number of other local worthies who put up a great deal of money for a building and a library. The Institution's first home was in Frederick Street, Shelton, now called Gitana Street after a local girl, Gertie Gitana, who achieved considerable success on the stage, having started her theatrical career as a diminutive gypsy in a charity concert-party. There is still a pub on the corner there called the Mechanics Inn. In 1859 the mechanics moved their institution to Pall Mall, building close to the British School and Art School, built in 1818. This is a sad and sorry-looking site now. The building had real architectural strength in a neo-classical sort of way; it had the obligatory Doric pilasters on either side of the double doors, and was one of the few buildings to have kept its contemporary cast-iron railings and gate pillars. The proportions of the building were so neat it had instant appeal, despite the addition of a third storey in 1880 and the rather forbidding barbotine portrails above the doorway of the extension built on the eastern end. These remarkable *tours de force* were done by a teacher from within, one George Cartlidge—another familiar name locally. What the British School gained the Mechanics Institution lost later on, its upper storey being taken down in 1956 when it became unsafe, leaving only the Tuscan first floor. The upper storey used to be Ionic. It was as nice a piece of architectural effrontery as you could wish for. The whole was occupied by the Library and Museum, which had been in various parts of the buildings from 1887 until their recent removals to new premises. There is no doubt that the whole mixed-up lot that made up the site was of interest. If subsidence and the bulldozer had not got it, then nature would have. It leaked at many of its venerable seams, grass grew on window ledges and drip sills, and willow herb inhabited the gutters and forecourt. It survived until the late summer of 1977.

Most of the old pot works have had to be demolished; they have served a useful life and Hanley's role is increasingly that of the shopping and commercial centre, with an increasing part to play in administration of one sort or another. New Law Courts have been erected in Regent Road, once an area of small manufacturies, low-grade housing and general dereliction. It is becoming increasingly smart and beginning to match the promise of the substantial late Victorian houses near the park.

Hanley Park was opened in 1897, and occupies some sixty-three acres of ground. Originally a large tract of waste ground called Stoke Fields, it was cut in two by the Cauldon Canal which was opened in 1777 to carry 'the navigation' across the Potteries, and which was indirectly responsible for the death of James Brindley. It was whilst he was surveying this branch of the Trent and Mersey that he was soaked to the skin, caught pneumonia and died.

The park has all the amenities expected of a respectable suburban park, and is a pleasant reflection of that spirit of generosity and public conscience that led the rich and influential, in Victoria's England, to lay out huge municipal gardens for the enjoyment of those who had no gardens because the same rich and influential people had crammed them like sardines into mean rows of terraced houses where a garden would have eroded someone's profits somewhere.

Not that the lack of gardens enclosed or imprisoned our Potteries folk. Many of them flew, vicariously, on borrowed wings. Countless back-yards had pigeon lofts and they flew their pigeons all over the country, some from as far away as Spain. We would do well not to underestimate the imagination of those whose frustrated genius bubbled and seethed in those teeming streets.

One old chap who admitted that the furthest he had been in his life was Etruria Station to send his baskets of birds off on their journeys, and who had lived all his life in a terrace off College Road (which used to be called Victoria Road), waxed lyrical when it came to the flights of his pigeons. He knew all the train timetables from about 1906, including the changes made and the delays to which they were susceptible. He had flown with those birds in his imagination. He described to me the fields and woods over which 'we' flew. He had the feeling of the wind between his fingers, the sun on his back and an irresistible impression of the soaring sensation as they climbed to find their bearings. He raced with them over the soot and grime, borne up above the smoke until distance drew an enchanted veil over the dereliction and changed the grim grin of industrial paralysis for the warm smile of the open fields.

Others have run on four feet over slag and cinder, their minds with their whippets; streaking free and fast in pursuit of a bundle of rags or in competition with other staved-ribs and pointed muzzles or just for the sheer exuberant joy of rushing along, accompanied by the cheers and threats of a kindred spirit chained, but only physically, by clay-heavy clogs and stiff limbs of a lifetime's climbing through deep tunnels, the black intestines of industry.

Man as a miner is an irritant to the earth's crust, and his constant probing to win the black gold has created huge heaps of steaming, reeking rocky vomit which clog ancient fields, diverting streams and paths as they grow to produce a skyline that rivals the pyramids of Egypt in size and symmetry if not in purpose. Even so, the pit heaps share an unintentional identity of purpose, for even if pyramids are not the tombs and monuments of famous men, the 'tips' are certainly the monuments to a nation's unsung heroes, and are often their only tombstone.

Every mining area had its crop of accidents, which on the whole accounted for more deaths and unhappiness than the great disasters which gained a wider notice. Thankfully increasingly modern methods in mining have reduced the rate of accidents in our pits, but every village and town in North Staffordshire has suffered some mass accident that lingers as a speechless horror, and serves to underline the quiet heroism of the collier. There is still a feeling amongst the public at large, even in the shadow of a tip, that between the major disasters, mining is a safe and steady occupation. This is just not the case. Ever since records have been kept, more underground victims have been claimed, in any given period, by other causes than by the explosions and inundations that catch the public fancy. The roof fall and collapsing wall have always killed more steadily than the great disaster, and few battles have ever claimed such a high percentage of casualties as our mines.

During the Crimean War, the actual battle casualties among privates was 2.1 per cent; miners were killed or maimed at the same period at the rate of 5.15 per cent. Miners got no medals and "as many curses as ha'pence".

Hanley's horizons are dominated by pit-heaps, mostly other people's, and one of these, nearer Burslem than Hanley, is the monstrous mound of waste spewed up by the Sneyd Pit. On New Year's Day 1942, Sneyd Pit exploded, killing fifty-seven men. The explosion was caused by coal-dust, and a nation fighting for its life on so many fronts took very little notice. One point is worth mentioning. During World War Two, mining was a reserved occupation; that meant that miners did not have to join the armed forces, and some were criticised for remaining at their jobs instead of joining-up. If it came to a toss-up between the two for danger, I should say it was a six and two-threes.

Two features are evident in the Potteries. Huge holes from which clay has been removed, called marl pits, and the aforementioned pit heaps. There is a great move nowadays to fill the pits with the heaps. This results in two areas of level ground where it was anything but level before.

Although local clay proved to be unsuitable for table-ware about two hundred years ago, it still had plenty of other uses. Bricks and tiles are fairly obvious users of coarse clay, but in the Potteries, clay for 'saggers' was one of their chief assets, and was derived from that incredible Black Band group of rocks. This superb geological liquorice allsort had yet another layer to help us. The minute animals and huge plants that lived on the hot, swampy plains of what is now North Stafford-shire, a couple of hundred million years ago, extracted most of the iron from the clay on which they thrived. Now iron is the ingredient in red clay that makes it unsuitable for high temperature firing because it reduces the melting point of the clay. The lower the melting point of the clay, the softer the ceramics that are made from it, and the more porous they are. Fine for bricks and flower pots, useless for tea cups and chamber pots. Thanks to the hearty appetites of the flora and fauna of the carboniferous era, we have some very high-melting-point clay, in layers batween the coal, the iron ore and the limestone. Because of its high melting point, the clay is called fire-clay. You can imagine how useful that is in an area that earns its living by using heat. Another name for it is saggar-marl. That is where we came in. Most people have heard of a saggar-maker's-bottom-knocker, but few, even in these enlightened days, know what one is. It is not a queer form of Potteries hedonist. It is a youth who knocks the bottoms of saggars to see if they are whole. You can try it with a tea cup. Hold it by the handle and give it a sharp tap. If it rings it is a good one. If it falls to bits it is no good and you must buy some more. The word saggar is one version of the word safe-guard, and describes the container, rather like a hat box, in which the ware used to be placed to protect it from the searing heat of the direct flame in the old bottle-ovens and to keep it from forming a permanent attachment to its neighbours. These clay boxes are no longer used by most potters, since the old type of kilns, called intermittent, have been replaced by continuous kilns in which the ware goes into a long tunnel on trolleys, through increasing and then decreasing levels of heat in an endless procession.

The saggars were once made by specialist craftsmen in the factory where they were to be used, or by firms who made nothing else. Now that they are no longer much needed, the holes can be filled up. Clay that is needed for the few saggars that are now used is obtained as a by-product of other processes.

Not that spoil-heaps and marl-holes are the only things that the people of the Towns would like to forget. There are the shard-rucks—that is pronounced 'shord-rooks'. A shard-ruck is a mountain or hole, full of broken bits of pottery and worn out plaster moulds (they are called 'mewds'). These two have been most lately used to fill holes of various sorts. Some have ended in quarries; some have helped to build motorways; others have gone into concrete; yet others have been used to obliterate canals and railway cuttings.

Now the people of the Six Towns do not do anything in a half-hearted fashion. In the sixties the Potteries had over two thousand acres of derelict land, no real slums, lots of new housing estates largely built on semi-agricultural land at the edges, and a hazy notion of unity that was not helped by the vast areas of dereliction and a crazy set of roads that seemed to go away from everywhere and got nowhere. At that time I was seriously assured by an aged lodge-keeper, who cycled everywhere, that Stoke was the only town in the world that was uphill, going and coming.

It may be as well to know what a lodge-keeper is. Often, but by no means always, an older gentleman with considerable experience of the potting trade, the lodge-keeper officiates at the entrance to a pot-bank—they are called pot-banks when you work there; they are "Refined earthenware manufactories" when they come up for sale—anyway, he is there to welcome and help visitors to the works, who are always welcome by appointment, and are often welcome unannounced. One of his jobs used to be to slam the gate in the faces of any luckless worker who came a microsecond late, and who then lost a day's wages. How that slam must have echoed through the 'tunnel'. Pot-banks were invariably entered through a tunnel, a kind of high archway piercing the centre of the building, with the board-room above it, and often a Venetian window. Everything that came and went did so through that tunnel, and usually still does, and past the lodge whilst it did it. The lodge is a sort of sentry box, often disguised as a room, but not to be mistaken for such. The only things that failed to go through the tunnel and past the lodge were those things which went over the wall at the back; still, many families have a black sheep here and there, but that is another story.

The lodge-keeper is no longer such an ogre. He is definitely not to be confused with the 'Gestapo' who guard some of our other types of factory, and the only thing that gets slammed now is a finger or two in "them new-fangled time clocks".

Lodge-keepers are not the only things that have changed. The most obvious change is in the landscape. Marl holes are now fields or ornamental lakes, pit heaps have vanished, there is no longer a pall of smoke over the valley; a brave new face is beginning to smile through moonscapes of the past.

Every Victorian town with any pretensions to self-respect built itself a park; the bigger the town, the larger the park, ranging from a few hundred square yards—a real park could not possibly be metric—to scores of acres as in Hanley. But now the pocket handkerchiefs with their neatly trimmed hems are a thing of the past; interesting museums of suburban niceness that remains nice but no longer help to recreate the modern family. Now we have a forest park. A combination of the hugeness of a forest with the landscapability of a park. Instead of gravel paths there are cinder tracks; no more are there borders of wall flowers, the old walls themselves flower with lichen, ferns and pretty little pendulous plants with blue or pink flowers that every one knows the name of but has just forgotten.

Instead of ignoring their sclerotic railway lines, their dug-out marl, their mined away coal, their tipped slag, and their blasted furnaces, the city planners looked to see what use was being made of them now. One thing was certain: if there was something there, the Potteries people would not be letting it go to waste. Where there was a path people walked, or ran, or rode or skipped. Dogs raced, ponies trotted, model aircraft shrieked. Cowboys defeated countless Indians. El Alamein was fought again and Scott still trudged intrepidly from pole to pole.

Where there was water fish were fished, swims were swum and boats boated. The King of Spain had his beard repeatedly singed, son of *Kon-Tiki* sailed again, frogs spawned and puppies were drowned.

The planners observed all this in an avuncular fashion and said to one another, "This wart is not a wart. This wart is a beauty spot." What industry had chewed over and spat out, the people had found was very much to their taste. The planners, who really were Uncles all the time, did what Uncles always do. They dug into their pockets and racked their brains to give us all a treat.

No praise is too high for what they achieved. Faced with apathy, suspicion, hostility and 'Trespassers will be persecuted' signs, they turned a sow's ear into a silk purse by the exercise of ingenuity, tact and cunning. Their assets were hard work, too little money and a handful of statutory powers that other authorities do not bother with.

In 1971 the Central Forest Park was opened on the site of Hanley Deep Pit. It covers what used to be eighty acres of derelict colliery and fifteen acres of old marl hole. Probably the biggest, nastiest mess in the area before Uncle Jim Plant and his boys moved in. The scheme won the top award in the National Conservation Awards Scheme. Trees were planted, pit heaps were smoothed and cooled to make gentle slopes for wandering paths to climb. At the entrance to the park stands the old winding wheel of the Deep Pit, a monument to the hard work and courage that ruined the place, and the same qualities that made it pleasant again. In the same year one of the city's most successful projects, the Westport Lake Park, was completed. This time a nautical Prime Minister came to open its forty-nine acres of sailing facilities and picnic areas. At the same time Mr Edward Heath launched the reclamation of the Glebe Colliery which had only closed in 1964.

Apart from anything else, these schemes were feats of engineering and skill unparalleled anywhere. New techniques were tried that have benefited large sections of Civil Engineering. Holes and mounds that were centuries in the growing were sculpted by mammoth machines in a few weeks. Not that all the reclamation has been given over to leisure areas. One of the tips that used to keep the midday sun off Hanley was Berry Hill. This was razed to make a new industrial estate. Whilst the monstrous mound was being carved away to fill many of the adjacent marl holes and ravines, the material was so hot that the tyres of the giant machines caught fire and the drivers had to wear special breathing equipment as the fumes were deadly and dense. Breughel would have loved it. No science-fiction writer could have predicted that scene. This is 'new lamps for old' in which the treasure has long gone from the cave and the cave itself becomes the treasure. The Genie wears a grey flannel suit.

In 1972 another huge area of despair and rusting relics was rescued by the Uncles when Monks-Neil Park was opened in Fegg Hayes. This one is a bit more formal than the others, but there are still no signs saying "Keep off the Grass". In the Potteries they lay turf and plant grass especially for people to walk on.

In March 1976 the Grange at Cobridge came into the news once more. Forty-nine of the ancient acres had been used for ages by all and sundry for the indiscriminate dumping of rubble, rubbish, household refuse and 'shraff'. Anything that the pottery industry wishes to discard is called 'shraff', and the place it is dumped is called a 'shraff-tip'. The distinction between a shard-ruck and a shraff-tip is rather a delicate one to define, but basically a shraff-tip is public and a shard-ruck is private.

The Grange Land Reclamation Scheme stretches from Etruria to Burslem and is one of the unifying agencies that are relentlessly joining the Towns together.

Another unifying force that has been at work for over 100 years is the *Sentinel* which Arnold Bennett called the *Signal*, and which published his first article when he was nineteen. The *Sentinel* was born into a period when the warring antics of rival newspapers made football crowds look like a parsonage tea party. Amongst the hopeful failures were the

Reporter, Spice, Potteries Daily Express, the *Staffordshire Daily Times,* the *Staffordshire Knot,* the *Weekly Star,* the *Staffordshire Courier,* and the *Staffordshire Evening Post,* to name but a few.

Naturally, the chief competition was from political rivals, and the most dogged opponent for the Radical, Liberal *Sentinel* was William Owen, a Labour leader; his organ was the *Knot* for which Arnold Bennett's solicitor father worked in an honorary legal capacity. He published it in the evening, in the morning, at midday, even weekly, but none of his ventures lasted very long, sometimes only weeks. It was this unfortunate journal that figured in Bennett's farce *The Card,* when rival newsboys fought in the streets. The finale of this episode was based, loosely, on fact. The *Sentinel* did have an annual 'treat' for its boys and in the story the rival newsboys, insanely jealous and slobbering Pavlov's dog to shame, were lured into a warehouse by the smell of toasting cheese and hot jam. Whatever you think of the patronising affectation of the rest of the story, the memory of that incident, recounted with rare genius, is always good for silent mirth. The book was based on the exploits, real and apocryphal, of Harold K. Hales, one of the best loved characters in the Six Towns.

The most tenacious adversary of the *Sentinel*'s monopoly was the *Evening Post* which represented the Tory Unionist opposition to the Liberal Home Rule policy advocated by the *Sentinel*. This ended with the two papers merging and a good deal of compromise in political comment. It was too much for the editor, Mr William Mackie, a Scots radical. He left and founded his own paper, the Liberal *North Staffordshire Herald;* it failed like all the others. As Bennett says in *Hilda Lessways*: "Nothing has ever stood up to the *Signal* yet".

The paper has a wide circulation and exerts much influence locally. It is regarded with respect and affection, the odd compositor's errors being treated as a sort of puzzle at which no one takes offence.

Whilst the news coverage is excellent, the paper is most valued for its advertisements, especially the small ads. Readers anxiously seeking to purchase a particular item secondhand, wait at the loading bay in Foundry Street to grab a copy hot from the press to be sure to be first on the doorstep.

Their premises are an interesting jumble. You can still make out the pigeon loft on the old 'New Offices' built in 1901, when news from local race events was flown back in serious rivalry for the new telephone exchange just a short way down Trinity Street. These offices were built on the site of the old, just as the present building, finished just in time for the Second World War, was tacked on to the front of its predecessor. One true story worth recounting is that in 1926, during the General Strike, all the machines had been closed down and the type metal had gone solid in the linotype machines. The Editor, sensing that an end to the strike was near, and wishing to publish the news when it happened, collected up the apprentices whose 'articles' prevented them from going on strike, and, having purchased every blow-lamp from a nearby ironmonger's shop, proceeded to melt the metal and prepare the new edition.

Whilst we are in Trinity Street, let us look at the old Telephone Buildings which are still there, dwarfed now by the new exchange built on the back in Marsh Street. The older building is very pleasant. Trinity Church is a nice bit of brick Gothic with a tower possessed of a small spire. A Gothic school room was later built on the uphill side of the church. The foundation stone of Trinity Church was half-obscured by the provision of a fine new porch and, in case the enigma upsets visitors, it was built in 1885. The school dates from 1912.

In Fountain Square, which is found by following Trinity Street eastwards through Miles Bank, there is a little collection of buildings which give a superb impression of the scale of Hanley in the early nineteenth century. There is the French Horn public house and the low, single-storey shops alongside; even the dust brick pavement is there. Moving due North from this unexpected bonus brings us to the Market Square, where a small plant market is still held. There is the famous Hanley Market, opened in 1849 on the site of the old Swan Inn, still remembered in Swan Passage which runs alongside. The market hall is now threatened with demolition as age and subsidence have taken their toll. The city can ill afford to lose the pleasant front of this building whose motif of vases and balustrades runs round the corner on the other shops. On the other hand, the population have derived their livelihood from the minerals of the area for hundreds of years, and can hardly escape the consequences now.

The consequences are not as hard for everyone as they were for Mr Thomas Holland. On 3rd December 1903, Tom Holland was walking along St John Street in the east of Hanley, minding his own business and carefully stepping between that peculiar bounty of the horse that was to make the motor car so welcome, when he must have stepped on a joint in the paving stones or something, because the ground opened up and swallowed him. Ropes were brought and lanterns were shone down the hole, but there was not a trace of the unfortunate Thomas. On 5th December, when all hope of his re-emerging was gone, a funeral service was said round the hole, into which a number of cartloads of shraff were subsequently tipped. Every parish has its story of holes that suddenly appeared, reputedly engulfing all manner of things, but the entombment of Mr Holland is the only authentic case that I can find. So prospective visitors need have no fear, provided they do not step on the joints in any paving stones.

Subsidence is a very gentle affair, and one does not awake to find gaping cracks suddenly opened up in the walls. They appear gradually, with a sort of shy indifference, like the growth of a favourite child.

And so has Hanley grown. The changes have been slow and, in the main, carefully considered. Where once stood houses and gardens, later were pot-banks, and now are bright new shops and offices. Hanley has the shops you would expect of a city of over quarter of a million inhabitants, and most of the nation's big names are represented. One thing they are short of is statues. For an area that produced so many out-standing persons, they are very slow to cultivate the cult of personality. When you think of the useless oafs whose bronze effigies ruin the outlook in so many of our towns, you would think that Hanley could manage some sort of memorial to Josiah Wedgwood, to whom they owe so much.

His youngest son, Thomas, has a statue in Etruria Park, which is behind St Matthew's Church, half of which had to be taken down a few years ago because it was falling apart. This Thomas was a leading scientist of his day, and his experiments contributed considerably to the early progress of photography. He got past the darkened bottle stage and produced impermanent images on cloth and white leather. What a family.

However, at Etruria, the Wedgwoods had, like Baron Frankenstein, created a monster they did not like and could not control, so Josiah the second took himself off to Maer, where the ancient Britons and the Romans had found sanctuary before him.

The monster grew and grew in the next century. Etruria's rustic charm was buried first under houses and pot-banks, and then under mines and iron works.

In 1839 Earl Granville built three furnaces at Shelton, to take advantage of the rich deposits of iron ore which he could smelt with coke made from the coal he was already mining. His son, the fifth Earl, erected new furnaces on the bank of the canal west of the Etruria works, and in 1852 he built a forge and rolling mills alongside them. That is where the area acquired its new name. In the past ten years or so the name Shelton Bar has stuck and it will certainly outlive the works. The Earl's new company, formed to run the mills and forges, was called the Shelton Bar Iron Company, and used the pig-iron produced in his blast furnaces. The rise in population in the 1850s, in Hanley and Shelton, is directly linked to the expansion of the iron works there, and their closure now would be a disaster.

Constant pressure from the mines and iron works first drove the Wedgwood family from their valley, and then their factory. Josiah Wedgwood and Sons started to build a factory in a garden at Barlaston, outside the city boundary to the south. The idea was much the same as had been Josiah I's.

New processes were being introduced and the old works could not be converted. Mining subsidence took a savage toll of the buildings. The housing that had been so modern in the 1770s was now virtually a slum. The canal was no longer the best way to move their goods or raw materials, having been superceded by the railway and Wedgwoods already owned a huge estate at Barlaston which was close to a main road and was served by both railway and canal.

Here the pattern could be repeated. An estate of new houses on modern lines, a new factory with all the latest gadgets, and transport ready to hand. The Second World War interfered with the development and by 1940 only the earthenware section was completed; the rest was done after the end of hostilities. In 1943, Wedgwoods sold out to the Shelton Iron, Steel and Coal Company, but continued to use space in the old works until the early sixties, having shared it with the Dunlop Rubber Company.

Now, in the 1970s, Wedgwoods is all at Barlaston, except that they now own a large number of other potteries who trade under their own famous names.

All that remains of the Etruria works is the enigmatic Round House, whose original purpose can only be guessed at, and, of course, the name.

Amongst the things that go to make up this capital of the Potteries is the North Staffordshire Polytechnic, reasonably enough in College Road. Currently running on two sites, one at Shelton and the other at Stafford, the Poly will move eventually to a new, single site right beside the baby Trent, in place of some allotment gardens and the old Hanley Sewage Works in Leek Road. This represents only the tip of the iceberg of educational concern that the city has. Everywhere there are colleges and schools, most of them new and bustling. Up College Road and opposite the larger portion of the park is the Cauldon College of Further Education, built on the site of Ridgways famous Caldon Place Potteries.

The oldest existing works, and the one with probably the longest history in Hanley is Ashworth's. These are to be found behind the Mitchell Memorial Theatre in Broad Street, but the huge chimney with its white 'Mason's' mark on Morley Street is hard to miss. To the industrial archaeologist this is a fascinating site. It was here that one of the greatest of our potters, John Astbury, had his works. This gentleman is said to have stolen the secrets of stoneware manufacture from the Brothers Elers, by pretending to be an idiot. They thought it would be safe to employ such a feeble-minded man, since he would be unable to understand, let alone make use of, their secrets. The secret, incidentally, seems to have consisted of refining the clay rather carefully, having first ensured that the clay was of good quality. John Astbury was responsible for the development of the oxide glazes on which Wedgwood later worked, and he manufactured a number of delightful pottery figures called 'toys', which were the forerunners of the Staffordshire figures and 'Flat Backs' that were produced in such vast numbers in Victoria's days.

Another 'spy' at the Elers' Bradwell works was Joshua Twyford who succeeded in gaining a working knowledge of their processes by a feigned indifference to what was going on. He was another Shelton man, and there is still a huge works in Shelton, between the canal and the railway in Shelton New Road, which bears his family name.

III

STOKE-UPON-TRENT

THE NAME is old English—Stoc—and merely means a place, usually where there was a church. That is what the central part of Stoke was until the nineteenth century: a church built near the confluence of the Fowlea Brook with the Trent. When Burslem was a thriving industrial town, in the 1760s, Stoke was the Church and the homes of the Rector, his curate and the parish clerk.

The fine, Perpendicular-style church was consecrated in 1830, and was the first of the new churches to be built in the Potteries in the nineteenth century. It was erected near the site of the original Norman church, the ground level being raised to avoid the flooding to which the area was prone.

In the churchyard two fragments of a stone cross were dug up in 1876 near the south wall of the old church. They are pre-Norman and are pieces of a Saxon preaching cross. They seem to have served as a door lintel at some stage in their history, and were placed in their present position in 1935 at the expense of Percy Adams of Woore Hall. The church also possesses a Saxon font and has many memorial tablets to great potters and their families.

In the churchyard are what appear to be the ruins of the previous church. There are two semi-circular arches, two piers, one circular and the other octagonal, some parts of other arches and assorted pieces of masonry. These are bits of the old church discovered in the mill-race of Boothen Mill when it was demolished in 1881. The Potteries antiquarian Charles Lynam discovered them and determined to use them as a memorial to the ancient centre of worship. Careful studies of old prints to which Mr Lynam did not have access, have shown that he should have made five arches from the fragments, not four. However, it is a good try contributing much to the atmosphere of the town, and it is always easy to be wise after the event. Local historians who criticise Lynam for his effort would be better occupied thanking him for bothering. It is one of the earliest instances of conservation on record.

A near contemporary of the Church is the Hide Market on the corner of Epworth Street and Hide Street. It used to look out over the old Town Hall, built in 1794, and its survival is as surprising as it is pleasant. Built as a covered market in 1835, it is brick with the obligatory stone dressing. Why could not the good people of this valley bear to see a piece of naked brick-work? There are arched entrances with iron grilles, guarding a pedimented central bay. The stone steps that once gave access to the main floor have been cut off at the knees, so it is a good job there is a back door.

Under the central bay is the entrance to the cellar which was once used as a jail and also as the fire station. The hall is unusually well suited to its purpose, very large and very cool. Stoke seems to have had an obsession for combining markets and town halls. The original market was in the old town hall, a little further along Hide Street on the same side, and was held on the ground floor with the meeting room above, quite adequate for the purposes of the inhabitants of the few scattered houses of late eighteenth-century Stoke. It closely resembled the 'town halls' of Market Drayton and Tamworth. The 1835 building was erected to counter the attractions of the market in rapidly growing Hanley.

By 1845, enough of the new Town Hall in Glebe Street was completed to allow the market to move there, into the ground floor of the central portion, with a cattle market in the yard behind.

The present Stoke Town Hall, far and away the most impressive public building in the Potteries, was started in 1834. It is, of course, of Classical design with a long stone front of thirteen bays and very tall, elegant windows. The first floor has a well-proportioned Ionic portico with its usual pediment. The three-storied, projecting side wings have a similar pediment. The ground floor is suitably rusticated with entrance through three round-headed arches in the centre. These arches are repeated in the flanking windows.

In 1883 the market was hived off once more to its present home in the Flemish style brick building in Church Street. This left enough room in the Town Hall for a mayor's parlour and offices, which used to be housed in the south wing.

This hall built of huge blocks of stone is the remarkable reminder of a doubtful starter who came in to win. Of the six pottery towns loosely articulated along the road running up the valley, Hanley was by far the largest and most prosperous in Edwardian days, having the greater powers of local government conferred by its charter of incorporation as a county borough and a very go-ahead administration.

Whilst all its neighbours were stuck in the past and behaving like isolated market towns, Hanley had its own school board, police force and quarter sessional courts; they even had a public swimming-bath. In every way they behaved like a modern industrial town, but, on their little built up islands of land, they were short of space. Thinking people in all the towns were aware of the stupidity and sheer wasteful-ness of unco-ordinated administration and development, and despised the jealousy and rivalry that existed between them. Each town, suspicious of its companions, was cautious in such basic things as way-leaves for sewage and drainage, the biggest problem of all. The colossal, wasteful cost of half-a-dozen different schemes of all kinds has a kind of pantomime quality that produces only hollow laughter when one thinks of the repeated mass burials of the victims of cholera and typhus —and this was England, not tropical Asia.

The progress of the towns towards responsible local government was usually along the same lines, from improvement or town-commissioners to boards of health, always a first priority. The Potteries people prayed for a cool summer; in hot, dry weather the stench was unbearable as the water-level dropped in its open sewers, the Fowlea Brook and Trent River. The chances of avoiding illness in a hot summer were remote for all but the well fed, and there were precious few of those.

Of the 'Big Four'—Hanley, Burslem, Longton and Stoke—Stoke was the slowest in achieving proper status. Longton was made a borough in 1865, Burslem made it in 1871, Stoke trailed in last in 1874; all well behind Hanley's triumph of 1857, which they would not have had but for their throwing in their lot with Shelton; an early amalgamation often goes unnoticed.

Tiny Tunstall and Forgotten Fenton became urban districts in 1894, at last giving each of the towns a recognised format in government and a civic head with some sort of parity of esteem with his colleagues.

None of which explains why, when the Federation of the Six Towns took place on 31st March 1910, it was called Stoke-on-Trent, instead of Hanley or whatever, nor why it became the administrative centre.

The reasons are largely historical, although I would like to think that there was a kind of prescience in the building of that enormous town hall way back in 1834, when it must have stuck out of the marshes alongside the new parish church like the twin fingers of derision in what was little more than a village.

The church had, of course, been the administrative centre throughout the Middle Ages, and custom may have played some part in the choice of Stoke as the seat of power for the new county borough. Then again, there was plenty of space. All the other towns had been built up early with the constant subdividing of the larger estates into ever diminishing holdings. The mud-flats of Stoke presented no such problems, and, in 1848, the last remaining bit of moor in the valley was taken over for the new railway station, immediately fixing Stoke as the new communication centre. She already had considerable wharf and warehouse facilities on the canal, and the hill up to Hartshill, known as Cliff-bank, gave scope for much livery activity. The turnpike had to go up that hill from the church and extra horses and men were needed to assist the wagons up the hill to the rich market of Newcastle. The railway brought even more business and in the 1880s, Stoke rapidly became an *entrepôt*.

The old town of Stoke had lain at the foot of Penkhull,

whose hill now holds North Staffordshire's enormous hospital complex high above the grime. Like most of the happenings in the area, the growth of Stoke was more concerned with the doings of two or three men than with any set of geographical or economic considerations.

The most important of these men to Stoke was Josiah Spode, or, more accurately, were the Josiahs Spode, since there were three of them. The first seems to have been born in 1733, and was apprenticed to Thomas Whieldon, across the river in Fenton, in 1749, a few years before Josiah Wedgwood's association there, and was working there as a journeyman when the other Josiah was in partnership with the firm. Spode left Whieldon in 1762 to be works manager for Turner and Banks at their factory in Stoke. When Turner and Banks broke up in 1770, Spode assumed control of the factory, and, as was usual at the time, a local solicitor went in with him with the cash. This was a Mr Tomlinson. A further association, this time with another potter, Thomas Mountford of Shelton, took place and the works, at the corner of Liverpool Road and Church Street opposite the Old Town Hall, prospered sufficiently for Josiah to achieve complete financial independence. He severed all other connections in 1776, and moved from the status of Master Potter to Owner. A creditable performance for one who started out as 'a humble workman'.

Much of the old works remains, hidden amongst the buildings of the present factory, a wall here and a floor or roof there, and as one drives round two sides of it in the gyrations of modern traffic, a chastening idea of the magnitude of the operation is acquired.

Josiah Spode I died in 1797, and Josiah II took over where his father left off. History is full of accounts of strong resolute fathers followed by weak, vacillating sons. This is not one of those stories. Spode's went from strength to strength. The two worked as partners when Josiah II's apprenticeship ended. He had been bound to his father in 1770, at sixteen as the elder had been to Whieldon. This had been when the Turner works was taken over. When the lad had learned the business he was sent to London to take charge of the warehouse and showroom, together with William Copeland, another local lad making good. It was under Spode the younger's influence that under-glaze decoration in blue transfer prints was pioneered in 1784 and it proved to be the making of them. Wedgwood's had disregarded this process as being not quite nice. Another bright idea that young Josiah had was to run his London warehouse as an emporium. The building was the theatre in Portugal Street, Lincoln's Inn Fields, where a South Staffordshire actor, David Garrick, had made his début. Under Mr Spode Junior's management, the firm sold glass as well as pottery, and they also sold porcelain purchased from other firms. For some time before old Josiah's death, they had been working on an improvement of the bone-ash porcelain originated at Stratford-le-Bow. At that time potters used one ton of clay mixture to five tons of coal, so the attraction of local coal, which burned with a long flame, useful for even kiln temperature, coupled with the strong tradition of pottery skills, had established the industry very firmly in North Staffordshire by the end of the eighteenth century. So, although much of the decoration was applied at London and other ports, the heavy work was done here, as was the expensive and painstaking development work, essential to any industry.

Up to 1790, Stoke-upon-Trent had not had a large scale industrial premises, neither had they seen a large house, such as the new industrial barons were building the length and breadth of Britain. Spode perfected his bone-china sometime before 1800. The material was a local imitation of true porcelain, and was made in very much the same way as traditional white stoneware, but with a higher proportion of glassy material and with the addition of calcined bone. It was upon this invention that the prosperity and further development of the potteries depended. The process was well known and established on a sound commercial basis by 1803, and Josiah Spode made an immense fortune from it. Bone china resembled true porcelain much more closely than any of the other imitations and, being very white, smooth and translucent, it gave an added clarity and beauty to the traditionally high standard of brightly coloured flower painting of the area. Spode remained well ahead of his colleagues in the profusion of patterns applied from prints, and the crispness of his blue-printed ware is evident on existing specimens even after the lapse of nearly two hundred years.

To accomplish all this, the Spode works had to expand, and Josiah II leap-frogged around the site, taking over houses, huts and hovels, until the factory extended to fourteen acres. He did not demolish the buildings acquired. We have one of those rare insights into the day and age of people long gone in that incredible heterogeny, the 1842 Children's Employment Commission. Thomas Shaw, said that he was the 'bailiff and foreman' of the works, and that they employed 454 men, 249 women and 77 children. A child was a boy or girl under thirteen years of age. He added that the 14 acres of land were all enclosed and consisted of 19 ovens, 272 working-rooms, 19 slip kilns, 42 warehouses, and 33 other offices.

Her Majesty's Assistant Commissioner confirmed the area covered and added his comments, quite unasked, on the state of the premises. They appeared, he said, to have been erected many years, as there was neither conformity, order or system of arrangement in any of the departments, except in the dipping-house, green house, and one or two others which were comparatively modern structures. The green house was where the 'green' or freshly made pots were set to dry out. He complained that the printing-rooms, throwing-rooms (where the pots were made on the wheel), pressing-rooms (where pots were made in moulds) and painting-rooms were close, low, small, and inconvenient places for the purposes to which they were applied. Further, that numbers of people worked close together and that even in the winter season, with "the thermometer at 38°F", they were hot and unhealthy. He could find no means of ventilation except at the doors or small opening at the windows, although there were hot-stoves and plates in every room. The hot-plates were to warm the shellac from which the transfer prints were made. His parting shot was that the premises were well drained. The premises had every reason to be well drained. Over a period of a few years, all the clay on the site would be dug out for use and replaced by shraff, which would build up the ground level with the cinders and ashes from the kilns. The kilns had to be rebuilt from time to time, and the bases, of bricks and pots and glaze, fused together by the immense heat of the firing, would be left and built over. Close by the works ran the Fowlea Brook, the Newcastle Canal and the Trent-Mersey Canal, into which this immense site, raised up on a platform of ceramic rubbish, could easily drain.

Spode's bright, cheerful blue-printed pottery brought tableware within the reach of the growing urban population of Britain, as Wedgwood had done for the rich a generation before. Not that Spode was the only firm to make the "highly esteemed blue-printed Staffordshire ware". There were scores of them at it, and many of them rivalled Spode for quality, but none for quantity as well.

When Josiah I had died in 1797, Josiah II took Josiah III into partnership with him, and also Mr Copeland, the tea merchant who had helped to run the London end of the business. The third Josiah Spode took little interest in the pottery. An early accident in those 'crowded mills' in which his arm was torn off cannot possibly have endeared him to the place; and since he did not have to work for his living—he had married a wealthy coal-and-iron heiress—he stayed at home. Home was The Mount, an extremely large and unlikely building of red brick with two huge domes. It was built in 1803 high up on Penkhull Hill, well away from the smoke and grime, and in a good position to watch the works. The building stands now hardly altered from the first and is used as a school for the deaf. Soon after taking over the business the third Josiah died and his wife and Josiah IV left the area. Mr Copeland's son then ran the business with his chief traveller, Mr Garratt, as Copeland and Garratt. Since then the family of Copeland has maintained their interest in the firm, and have been active in everything going on in the area.

It was during the time of Copeland and Garratt at Stoke that the Potteries' Strikes of 1825 and 1834–1837 took place. These followed a period of general unrest, during which most of the Government controls over trade unions and combinations of workmen had been removed. Now that they were allowed to unite to attempt to achieve a change in their conditions of work, they sought very mild and limited

objectives. They wanted an increase in piece-rates; the aboli-
tion of the 'truck system' which forced workpeople to buy at
company shops, always at a disadvantage and often by
payment solely with tokens; and the regulation of the sizes of
wares.

The operatives were not alone in wanting the truck system
outlawed; small manufacturers considered it unfair competition when the 'big boys' could offset a large part of the costs
of production by paying in 'truck'.

Local shopkeepers did not like it much either. The opening
of a truck store on a nearby pottery, usually the only
employer within easy walking distance of the area served by
the shops, had much the same effect as the opening of a
supermarket in a village today. The fact that the system
flourished at all is a tribute to the power of the manufacturers
if nothing else.

The earlier strike of 1825 had realised the object of
removing the truck shops, but no standardisation of sizes of
pots had been made and wages went down, not up. It is very
hard to get a clear view of what was happening from the
bitter and vituperative publications of the day.

John Ward, who was a local solicitor, writing his *History of
the Borough of Stoke-upon-Trent* seven years after the strike, said
that the operatives systematically 'turned out' in a mass from
any pottery where their prices and rules were not granted, and
that the unemployed were supported by weekly allowances
from those who were in work. This last is so indignant. It
comes as a shock to us to find anyone annoyed at the poor
helping the poor. It just was not cricket. He says that the
activities of the workers "greatly inconvenienced the Masters".
What is most intriguing is that the demands of the 'hands'
were so moderate, and that they made no attempt to improve
their working conditions which were often appalling. Walking
past the actual works and rooms in which those people
suffered is, to the sensitive, a weird experience. Stoke has
retained much of its nineteenth-century atmosphere, and it is
not surprising to find that all the old works has a ghost of one
sort or another. You have only to ask.

In the early days, when Stoke was just getting started,

Samuel Spode, one of Josiah's sons who had a nice new pottery at the Foley—we will come to that later on—sacked or suspended some of his journeymen. He had even conspired to have four of them incarcerated in Stafford Gaol

> because they would not submit like slaves to his imperious authority in going to dinner at an hour when neither Mr Turner's Men, nor any Men about the neighbourhood go, except one Work which is of little account near the Toll-Bar, and tho' they have offered their Services, desiring to work Peacably, observing the same Rules with their neighbours, he has been so far from listening to the distress of Poor Men who have got Families, that he has absolutely toss'd their Clothes into the Lane and trod them in the dirt, threatening them if they came upon the Bank.

So went the Hand Bill published at Lane End, 14th April 1791. Despite the seriousness of the pamphleteer, it is hard to escape the ludicrous humour of the situation. The self-righteous Mr Sammy Spode jumping up and down on the tattered rags his workers called clothes, spattering his high-gloss boots and doeskin knickers with pit-muck, whilst the starving men are led off in chains to Stafford Gaol.

To be quite fair to young Sam, he had got together with a number of his mates to decide on when the workers should have their lunch-break, and they agreed, as Master Potters, that a more equal division of the day would be arrived at by breaking at one o'clock instead of noon. Then, when some of their men murmured, they, the Masters, backed down and changed back to 'noons', leaving Sam Spode in the position of the lone Lancer, whose troopers have found something more important to do half-way through a charge, and left him leading very much from the front. Being a man of principle he stuck to the agreement, though for how long we do not know because the outcome of it is not disclosed. He ended his counter-blast with: "We are on all Occasions ready to promote the interest, Welfare, and Prosperity, of our Work-men in every instance, and trust that in forming this Regulation we cannot be called their Enemies."

How well the "interest, Welfare and Prosperity" of some workers was promoted may be judged from an account given by a forty-one year old scourer; that is one who rubs the grit and flinty particles from the newly-fired ware. She said she had worked in the pottery for twenty years, four of them as a scourer and had one daughter who worked at the same premises. This woman, Annie Williams, was in charge of two rooms and had six persons working with her, none of them children. She said that they received the ware from the ovens in a rough and dusty state, and that their job was to remove any bits that had stuck to ware during firing by rubbing them vigorously with sand paper or on grindstones. Whilst it is true that none of the twenty-or-so 'gentlemen' who signed the *Notice to The Public* said that they would take care of the health of their workers, our Annie said that scourers' work was very unhealthy, as it "stuffs a person up very much in the stomach", and that not many scourers lived long. It took some off sooner than others. None of them was ill, she said, except that they all felt overloaded upon the chest, which made them cough very much, especially in the morning when they first began. These poor lassies were paid by the piece and had, on average, eight or nine shillings dropped into their skinless, bleeding fingers, for a week of sixty lung-rasping, nail-rending tormented hours.

Even though daily doses of human suffering and indignity presented in films and television, as 'entertainment' or news have made such events almost commonplace to us, the resigned simplicity of the personal statements of pottery workers of the 'good old days' makes them too harrowing to read very often.

Annie Williams was working at Minton's, a nearly new and highly esteemed manufactory. Their scouring rooms were spacious and well ventilated, according to the Commissioner. He noticed that even there particles of dust were flying everywhere; the air was loaded with it. That was in a new china factory. What of conditions in one of the rambling old pot-works? Well, one of the most rambling and oldest pot-works in Stoke then was Spode's—at that time called Copeland and Garratt's.

A scourer there had worked in less hazardous parts of the works until she had lost her place whilst she had her family.

One of the real evils of the day, not only in the Potteries but in all other industrial areas as well, was that the manufacturers would rather employ a woman cheaply than a man at greater cost. The girl at Copeland and Garratt's, Sarah Henderson, was forced into the scouring room to feed the surviving child of the three she had borne because her husband, a skilled potter, could only get odd jobs. She was so 'stuffed-up' in her chest she could not lie down at night; her throat was always sore, causing a constant cough and it was difficult to breathe. She never had any medical advice—it was no use while she was at work there.

Mr Scriven, the commissioner—what a splendid Dickensian name for a clerk in the middle of a real bit of Dickens country—Mr Scriven noted:

"This woman's voice is scarcely audible. She is suffering in common with many others at this work."

However, the sleeping giant of public opinion was stirring slowly. In the 1830s Sir Robert Peel, a convert to Staffordshire, infuriated a lot of his mates by supporting reforms of all sorts, which was a step in the right direction, albeit a hesitant step. And the lovely poems of Blake and Clare were a sign of the times, serving to enlighten and stir up the middle class who had just acquired the right to vote—most of the men, anyway.

Although it was a mini-revolution in France that persuaded the British Government to enact a number of reforms, there were other forces at work. Numbers of sanctimonious, smug, self-satisfied books were being written for the sons and daughters of gentlefolk. Most of these obnoxious little volumes followed the current vogue for prints of George Morland's paintings of the previous generation, which showed how nice it used to be to be a farm labourer—if you happened to be one of the lucky ones who had not been booted off your hard-won tilth.

Some of the others tried to show, in a soppy sort of sentimental style, how indebted we all are to the poor people who have the misfortune to have to work for their livings. One of these, called *Little Jack of All Trades; or Mechanical Arts Described* (suited to the capacities of children!), was published in 1823. The author, who cunningly remained anonymous, but whom I suspect to have been a curate's wife, had gone to a lot of trouble to describe a large number of 'trades'. She probably had her information from one of the encyclopaedias which were then being issued in weekly parts. She had certainly never been on a pot-bank, although there were plenty in London she could have visited. To get the full flavour of the age it is a good idea to read the 'verse' that heads the chapter on 'The Potter', but anyone with a weak stomach can skip it and head for the text.

> Sweetest ladies, I beg, when you're over your tea,
> Admiring your china, you'll think of poor me.
> My China with Dresden or India will vie;
> Then why sail to Canton to get a supply?
> With Dame's corner cupboard can China compare,
> Tho' all is right English, true Staffordshire ware.
> Little cups, little saucers, and little cream-pot,
> Because she is good, little Mary has got;
> This ev'ning she may a small party invite,
> Her tea-table set,—what a beautiful sight!

She goes on, in the text, to confuse the potter's wheel with the lathe and bone china with salt-glazed earthenware, and concludes with:—

In viewing any article which contributes to our pleasure or domestic advantage, we should enquire of ourselves whether it has caused great inconvenience and painful confinement to the maker, or been the cause of shortening his life one moment. The manufacture of China is not exempt from pernicious consequences; for the process of gilding is extremely hurtful to the health of the workmen: therefore, when we examine this elegant ware, though we may admire the fineness of the composition, the brightness of the varnish, the lustre of the gilding, and the exquisite delicacy and beauty of the painting and the colours, our pleasure and admiration are not unmixed with regret.

Her heart is in the right place, but with information as confused as that, how could they have hoped to get their social sums to come right? In the decorating of china, gilding is not all that harmful. It was the poor silversmiths who stood to suffer from gilding. They dissolved gold dust in mercury, painted it on their silverware and then heated it to redness to evaporate the mercury, leaving a thin layer of gold clinging to the surface of their work. It would be hard to find a more poisonous substance than mercury vapour. In ancient civilisations it was the sort of work that would have been done by condemned criminals. The silver gilders were condemned all right, but they were probably not criminals.

In the potteries at that date, gilding was done by mixing gold dust with ground up glaze and painting it on the pot. The grinding up of the glaze was a killer, but the painting on of the gold was one of the better jobs. After firing, the mixture of glaze and gold was burnished to reveal the gold underneath. There were other ways of gilding used later on, but none of them was harmful to the gilder.

Not that the painters and gilders lived a life of idyllic industrial comfort. Most of them were girls who started work at eleven or twelve years of age, embarking on an apprenticeship of four years. In their first year they had one shilling a week for a nine-hour, six-day week, rising through one and sixpence in their second year, two shillings in their third and halfprice of women's work in the fourth year. If they had then become fully proficient—and the standard was unbelievably high—they went on to full wages, when they allowed the 'master' of the painters and gilders fourpence in every shilling they earned for the privilege of being permitted to work. The places in which they turned out work that was normally good, often exquisite, were dark, dirty and as hot as the rest of the works. Even after the glazes ceased to contain dangerous levels of lead, the paints often contained plenty—and copper too for good measure. The paintresses rarely had anywhere to wash and ate where they worked. They were frequently fated to premature blindness, were lucky to avoid consumption and, since they sucked their brushes to get a good point thus were subject to all manner of cramps and fluxes. They may well have wondered why.

Apart from the strikes which were as fashionable a way of airing grievances then as now, the way out of the miseries attendant on industrial progress in the ceramic industry was sought in a way that was becoming popular amongst nations, but which seems to be unique as an industrial manœuvre. Emigration.

In 1844 the Potters Joint Stock Emigration Society and Savings Fund was registered with the appropriate authority, as required by Act of Parliament. This Society was the brainchild of William Evans, who, as a painter and gilder who had served a full apprenticeship, had seen all the evils rife in North Staffordshire industry and had lived through the disastrous strike of 1836–7. Like a lot of others he was unable to find any employer to 'set him on' when the strike was over, and he, in common with many of his fellows, sought a place in Worcester, a town which at one time threatened to rival the Fowlea Valley as a centre of the china trade.

In 1838, Billie Evans was back in the Six Towns where he founded yet another newspaper, the *Potter's Examiner and Workman's Advocate*. His idea was to improve conditions in the industry by providing a means for people to air their grievances.

There were so many things wrong that it is difficult to pick out the main ones, but they seem to have been the annual contracts which bound a potter to his master, but not the other way round; the practice of the owners of only paying for those wares that came 'good-from-the-oven', and the 'allowance' system where the adult workers allowed so much in the shilling to their employer—who may or may not have been the owner. The 'hands' were not too keen on machines either.

Anyway, the *Potter's Examiner* shamed the owners into ending the 'allowances', and that effectively put up wages, but the physical conditions were as bad as ever. And although the masters agreed to pay a certain sum—it varied according to the type of work—for every dozen pieces worked on by each operative, what is hard to grasp is that, if you were making soup tureens they might be only ten to the dozen, whilst egg cups could be twenty-five to the dozen. A dozen was only twelve if you thought it was. So, at one works you could be making twelve of something for a penny while at another you would have to make, perhaps, nineteen.

Nineteen to the dozen? That sounds familiar. It also worked the other way round, of course. In these days of decimalised pre-packing and standardisation the idea that a dozen eggs, in time of plenty, could stay at the same price but vary in number upwards is quite foreign, but it was perfectly understood in the olden days. The hideous consequences of such an attractive sounding phrase as "nineteen to the dozen" were not really apparent. Until we examine a group of towns who count their produce in thousands of dozens daily. How many could rarely be found by multiplying the number of thousands by twelve.

So the editor of the *Examiner* was not short of copy. The more he ranged about the Towns, the more he realised that somewhere a break must be made into the truly vicious circle of working life in the Towns. His paper canvassed members for the emigration society. They paid one penny a day to belong to the scheme. Daily payment was important because not many employees worked a full week. Many of them did not want to; drunkenness was cheaply achieved as a simple anaesthetic for an intolerable life, and a sore head and a steady hand do not go very well together. Workers were often given no work to do when they went to work, although many would hang around for most of the morning hoping to be 'set on'. The real tyrants were the old bottle ovens, which are known technically as intermittent kilns, and much of the stop-go nature of work was due to the way the kilns were used or fired.

Since there seemed to be no way out locally, consolation was sought in escape to the New World. They would exchange a life of industrial hell for one of pastoral peace. Once sufficient funds were available a large area of land was bought in Wisconsin by their agent, Thomas Twiggs, who divided the land into twenty-acre farms, erected suitable buildings and sat back to await results. An epidemic of America-fever swept the Towns, and the enthusiasm engendered by the scheme raised morale considerably. Children, streets and pubs were named America. Only one America Street remains, in Tunstall, and the American Hotel survives in Burslem.

When there was fifty pounds in the kitty a draw was made, every member's name going into the hat. In 1846 Evans had accumulated enough cash to send eight families to paradise. The lottery was held in Hanley Covered Market and was made an occasion for a huge bun-fight and concert. A great deal of corporate effort went into the send-off. Friends and relatives gave clothes and tools to the lucky ones, and when they finally left Etruria, by barge, for Liverpool and freedom, musical bands followed them for miles along the tow-path, and thousands of cheering spectators lined the banks.

After the first flush of ill-informed enthusiasm it all went quiet. Many of the settlers found the change from home too great to bear and returned, beaten. Some of the others neglected their farms to go hunting and fishing. Very few managed to make the idea work. The town that grew up amongst their farms is now called Pottersville. One of the things they do not make there is pottery.

The stories of hardship and privation brought back by the returning settlers dampened the enthusiasm of the potters. The scheme slowly faded into oblivion between 1847 and 1849, and the union, which had poured its funds into the venture, was broken, leaving only small weak unions representing individual trades, rather than one powerful union for all. For the sake of a happy ending to part of the saga, Evans, who had left the Potteries after the collapse, to work as a reporter in his native Wales, was invited back to the Swan Inn, Burslem, in April, 1854. There he was presented with an illuminated address thanking him for his efforts and a silken purse containing twenty pounds. That was one 'sow's ear' that turned out all right.

On 22nd August 1844, the *Examiner* published, in the form of a letter from a member of the Black Lion Lodge, the Ten Commandments of the union. Number eight read:

Thou shalt not introduce machinery, nor any invention that will supercede manual labour; for surely as thou dost, evil will come upon thee.

This echoed the war-cry of the contemporary working-class movements in other areas, where manual labourers who had been deprived of their livelihood—or feared that they would be—by newly invented machines, had broken out in a spate of wrecking and burning, with here and there a little mild pillage. Two years before the latter-day Moses had come upon his new set of Industrial Tablets in the burning hills of North Staffordshire, there had been some pretty routine machine-bashing and rioting in the area, accompanied by the obligatory window-breaking and rick-burning at the premises of Local Government officials.

The scene of one of these incidents was a house then called Beech Grove, in that village-amongst-the-towns, Penkhull. It was the home of the stipendiary magistrate whose servants heroically saved the family silver by hiding it in a pig trough, when the waves of Chartist violence initiated by the Leicester orator, Cooper, reached their high-water mark. On their way through from Crown Bank, Hanley, the mob had wrecked the engines at Earl Granville's works, broken into Hanley Post Office, ransacked a pawnbroker's, turned over the house of the local Collector of Taxes and then rampaged on through Stoke where they smashed the windows in the police station, a fate shared by Fenton and Longton bobbies. They rifled the rectory at Longton and then set fire to it, looted a large house at Heron Cross and another at Great Fenton, ending up with Mr Rose, the Stipendiary at Penkhull.

They carried on in much the same way the next day, 16th August, the anniversary of the Peterloo Massacre, and were joined by textile workers from Macclesfield and Congleton. Estimates put the crowd as being between 6,000 and 8,000, having collected more spinners and weavers in Leek. They joined forces outside the Big House in Burslem—there hardly seems room—where they were met by a company of Dragoon Guards and local Volunteers.

The Riot Act was read, some stones were thrown and the soldiers fired upon the mob. One man was killed instantly, another lay dying in the street and many wounded were carried away by their mates. Some died later. The troops pursued and captured 274 of the rioters; their trials lasted a fortnight. The sentences varied, and give an interesting insight into the ideas of crime and punishment then in vogue. The ringleaders were sentenced to transportation: eleven for life; thirteen for twenty-one years, nine for fifteen years, eighteen for ten years and three for seven years. One hundred and forty-six were sentenced to various terms of hard labour. Eight were imprisoned without hard labour, fifty-five were acquitted, six were discharged by proclamation, two were held over to the next assize, two were fined and one released on bail.

It is doubtful that Sir W. Follett, Solicitor-General, who arranged all this, was aware of the conditions that roused the men to such desperate acts. Little by little the public became aware of how the other half lived, and in the same measure the operatives became accustomed to machinery. Paradoxically, the industry did not shrink, neither did the labour force. Both expanded although there were still rough patches, and hard times lay ahead.

It is all so peaceful now. Where frightened young soldiers stood unflinching to be stoned, and starving, coughing potters were shot dead in their tracks is now a bandstand. The gardens where pig-swill concealed the family treasures of Mr Rose now ring with the happy chatter of well-fed children. Beech Grove has become the Grove School. It is hard to believe it ever happened.

Although it suffered some drastic demolition in the early 1970s, Penkhull retains much of its quiet, village character and it is easy to see why so many local worthies in Victoria's day chose to build their homes there. The tell-tale signs of sub-sidence are largely missing, and what little there is seems to be the product of nineteenth-century jerry-building, rather than mining. Indeed, the extractive industries seem to have confined themselves to quarrying some of the excellent local stone, some of which was used to build the new church at Hartshill in 1842.

Despite the great antiquity of the settlement (and under that lot up there is a prehistoric hill-fort) the lovely little church in the middle of its triangular green dates only from the nineteenth century. Of far greater antiquity is the Greyhound Inn. It was here that the Manorial Court of Newcastle-under-Lyme sat. What is visible from the outside now is largely 1930s, but underneath is a fair proportion of the old timber frame, dating from the times of Good Queen Bess, as is the enormous stone chimney. Inside is some panelling, also of Armada date, probably from the courtroom itself. Mention of nearby Newcastle is a reminder of how the history of one area spills over into another, blurring the boundaries if nothing else. The road down the hill to the west leads into Newcastle, through Spring Fields and up to the Lyme Brook, one of the few natural boundaries in North Staffordshire. Hidden under the tarmac and concrete of Spring Fields and Trent Vale towards Hanford is a Roman Fort. In the area of rising ground above the Trent in Trent Vale was a Roman pottery works. Exhibits from the site are on show in the City Museum and are worth a visit, like so many other exhibits in their priceless collections. The Roman pots were made close to the present tile works. Whilst collecting ridge tiles from there for my roof, I enquired if they were still digging their own clay on the site. "Oh yes! From the same hole as Julius Caesar." He was about 100 years out in his Roman rulers but I knew what he meant.

Potteries people have a well-developed sense of history, though the view is sometimes a little hazy. For instance, everyone knows that, during the Napoleonic Wars, an organ grinder's monkey was hanged from a lamp post as a French spy, since it was common knowledge that French spies were wizened little men in fancy waistcoats. What they cannot tell you is which town was host to the unedifying spectacle, only that they are sure it was here. Some other towns have a similar story. I expect it is apocryphal.

While considering the taste for the traditional that many of the citizens of the Towns have—with three Labour M.P.s I hardly dare call them conservative, however small I make the 'c'—one of the things that crops up with great frequency in old records is the fact that people were hired from Martinmas to Martinmas. 'Martlemass', they usually call it. Now, Martinmas was 11th November, the traditional end of the agricultural year. This was the day when all the swine, cattle, sheep, guinea pigs and so on were slaughtered to be cured for the winter. Excepting two of each, like Noah, of course. In Tusser's *Five Hundred Points of Husbandry*, which he wrote in 1744, he says:—

> When Easter comes, who knows not than,
> That veale and bacon are the man?
> And Martilmass Beefe doth bear good tacke,
> When countrey folke do dainties lacke.

He has a note: Martlemass beef is beef dried in the chimney, as bacon, and is so called because it was usual to kill the beef for this provision on the Feast of St Martin, 11th November. So, now we know. But if we look more closely at the hiring arrangements, we see that they were hired from Martinmas to Martinmas Eve next. That is a year all but a day. The practice has roots that go back a long way. I have already shown that Queen Elizabeth made a mess of the local arithmetic when she divided our six towns up into five, in order that the poor might be cared for. Her Act was incredibly good for its time, but over the years it had gathered a lot of moss. It was passed in 1601 and the idea was that the ratepayers of a parish were responsible for their own poor. You could be officially poor for only three reasons: you were a young orphan; you were too old or too ill to work, or both; you were able-bodied and could not find work.

The ways of dealing with the categories, as time went by, became universal, very interesting and hideously inhuman. If you were a young orphan then an 'apprenticeship' was found a long way off where no one would ask any questions. If you were old or ill you probably died and that solved that little problem. If you were an able-bodied man and out of work, you were obviously a lazy scoundrel and so they built a special place for you to work in where a job was provided that was too close to sheer torture to bear close scrutiny. The job —usually picking oakum—combined with the degradation of public 'cleansing' and ill-fitting, musical comedy type prison clothes, were enough to make you wish either that you had never been born, or that you had found a job anyway.

There were exceptions, of course, where the spirit implicit in the Poor Law was carried out humanely and well, but those exceptions were very few and far between, and were more often in small village communities than in the growing industrial towns.

This may seem to be a long way removed from the hiring of a saggar-maker's bottom-knocker, but it is right on target. People have very involved minds. Just watch closely, till truth makes all things plain. Everyone was the responsibility of one parish or another. The parish that was responsible was the place in which a person had a 'settlement'. There were various ways of acquiring a settlement. You were settled in a parish if you were born there, if your husband or, if under sixteen, your father was born there; or—and this is the one that mattered most—if you had lived there for one whole year, or worked there for one whole year.

Since the manufacturers were ratepayers and had to find the money for the poor for whom they were responsible, they did not want to give anyone a settlement by employing them for a whole year. So everyone was fired the day before the year was up and, all things being equal, they were re-hired the day after so that they never became a charge upon the rates. As the bulk of the people who came into the area to make the industry possible were farm labourers 'on the tramp', who had been used to having Martlemass off to slaughter and cure their animals, they continued to have the same day off in which to get hired and fired and so retained their settlements in their home villages, which is why a lot of them died of neglect, and why, indirectly, there is a pub in Burslem called the American and why too many old people nowadays would rather die than ask for help. The place set aside for the officially poor, called paupers, was the workhouse. They went out with the onset of the National Health Service, and were, latterly, generally very good geriatric hospitals, but the shadow of the workhouse still spreads fear into the hearts of old people whose traditions have long roots.

Stoke-upon-Trent's workhouse lies right by the boundary with Newcastle, north of Spring Fields on the London Road, forming part of the City General Hospital. From the outside the dark, three-storied blocks are still rather sombre and oppressive, standing like the barracks of some great fortress at the foot of Penkhull. Inside there is all the bright, calm confidence of a great modern hospital. They have an enviously high reputation for care and comfort.

In the 1950s I was teaching in a school just across the border and was sometimes brought a skull or a handful of bones from the Paupers' Burial Ground. This was on the other side of the road from the City General, behind a low wall, amongst some trees beside the old Newcastle Branch Canal. In the early days of the workhouse, unclaimed paupers' bodies were taken over the road in their night-shirts and buried with a minimum of formality, beneath a thin scrape of soil. The activities of terrier dogs, small boys and erosion sometimes brought some of these pathetic remains to light.

Today the old workhouse site has been covered with a large number of buildings of all periods from 1842 down to the present. Right in the middle is one of a pair of the most attractive buildings you could wish to find. A perfectly proportioned, two-storied, Regency-style building; it proudly proclaims its erection in 1842 as a hospital. The plaque which carries this information is really beautifully lettered. It is a tribute to the good taste and historical sense of the local medical authorities that it is suffered to remain, unaltered, as part of the unimaginable complexity of modern medical administration.

The canal alongside which the paupers were buried has been filled in gradually since 1921, but parts of its course are still to be seen here and there. From opposite the Workhouse it followed a leisurely course southwards along the valley of the Lyme Brook until it had cleared the high ground that separates the two towns; there it turned east, crossing the A34 at Trent Vale to pass right by the site of the Roman Pottery Works near Rookery Lane. The Newcastle Cut then turned north just before the Trent accepts the doubtful blessings of the Chitlings Brook. The dried-up bed then passes the Michelin Tyre Factory, and forms the basis of a long, narrow garden from there almost as far as the Villas, which lie between the ancient hamlet of Boothen and Stokeville which lies at the foot of Penkhull.

The Villas were built about 1850 as a very early form of middle-class suburb. The best way of keeping up with the

Jones's in those days was to go on the Grand Tour, an agonising, dangerous and insanitary tour of Europe. During the journey, the terrified tourists made sketches of all that they saw; they observed the quaint customs of the natives and vowed never again to leave England, Home and Beauty. The up and coming young potters had little taste for foreign travel at that time, Llandudno being far enough for most of them. But they were not deprived of an insight into what it was like to live in Italy and so forth. There were plenty of Curtain Lectures at the various Sunday Schools and Chapel and Church functions, all augmented by lantern slides, hand painted at first, but genuine photographs later in the century. The physical reality was amply provided by a stroll round those villas. Very decidedly Italian in style, they have flattish, pantiled roofs, overhanging eaves, round headed windows and belltowers. To complete the illusion, a number of the continental ceramic workers imported by Mintons in the 1860s lived there.

Minton's main factory is a little closer to Stoke, and again, was built alongside the old Newcastle Canal. The founder of the firm, Thomas Minton, who must take his place as an equal with the Josiahs Spode and Wedgwood, was not a local lad making good. He was not even a potter to begin with. He was a Shrewsbury man who became an engraver. Apprenticed originally at the Caughley China Works in Shropshire, he went first to London where he did some engraving for Spodes, and after his marriage moved up to Stoke to work as a free-lance designer and engraver. Many of his patterns continue to be used, more or less modified, up to the present day.

Brief acquaintance with the industry at its heart served to show Thomas that, whilst there was not much to choose between the actual ware of rival potters, it was the decoration, especially the under-glaze blue prints that he was doing, that sold the goods.

Four years or so after arriving in the area, he bought the land where the factory now stands and, determined to benefit in greater measure from his own genius and skill, he started to build a factory of his own.

Like Wedgwood before him, he had the help of a Liverpool

merchant, William Pownall this time. He also had the services of the brothers Joseph and Samuel Poulson on the potting side. They owned the works adjacent to Minton's land, living on the site of the house from which the factory had grown. Very little is known of the ware made by these brothers before they joined Minton's, but, Joseph, the potter, and Samuel, the modeller and mould-maker, must get much of the credit for the superlatively high quality that has always been a feature of Minton's products. Their recognition locally was marked by the naming of Poulson Street, just up the hill from the works. Most of the street has recently disappeared in a lather of redevelopment.

Minton's however, is still very much with us, and their name has always been linked with all that is best in English Bone China, although, for a time, between 1816 and 1824 they switched production to earthenware and Thomas Minton started with the ordinary white and cream-coloured Staffordshire Blue printed ware. That got them off to a good start and a year after opening they introduced their china.

Since the break after Napoleon's wars, they have been into everything ceramic: always in high style and fine quality. I have often heard good things from the Potteries described as "Very near as good as Mintons". In the Golden Age of English china in the middle of Victoria's reign Mintons attracted skilled and prominent workers from all over the place. Spodes suffered by the loss of some of their best operatives, and painters and gilders from Derby and Worcester were enticed by the prospect of joining in an enterprise that emphasised quality in all departments.

At the time of the Royal Commission, they were trading as Minton and Boyle, and although the processes there were as deadly as anywhere else, the commissioners emphasised that the conditions were the best possible. The firm made every attempt to encourage their young workers to attend school, and various members of the family and firm contributed to local educational schemes. The area of Stoke near the Minton factory has benefited scenically from their association.

The factory itself is an imposing and pleasant building, far and away the most impressive industrial building in the town.

Recent demolition of adjacent properties and road widening have revealed its very long, three-storied front, with its four-storied central section in the way that its architect must have planned. Close to the modern addition to the bank is one of the Potteries rarities—a statue. It is an effigy of Colin Minton Campbell who presented the hilariously Gothic drinking fountain on the other side of the road in 1859. The fountain has stayed there since, but the statue has been moved about a good deal until coming home, as it were, to the factory.

Also Gothic in style is the College of Art, built in 1858, encouraged by Mintons, opposite. It has terracotta bands and window heads, or, if you prefer, "lines of moulded brick and posh lintels", as it was described by a near neighbour.

Far from Gothic is the library-cum-museum next door. It must owe something in inspiration to the villas, down the road. The design has been described as "original". Who could argue with that? It has enormously wide eaves and the main floor has great, round windows. There are also some very attractive mosaic and tile panels. I should not be a bit surprised if a cuckoo the size of a donkey came out of one of the portholes and yodelled the time. Mintons gave the land.

Altogether, though, the whole thing—Mintons, the Library, the Art School, the statue—make an attractive corner in a town that has progressively mislaid such corners.

Whilst in the area of Boothen, one can hardly avoid noticing Stoke City Football Club. Most of the infantile graffiti on public and other building has to do with the team and opinions about it. Like any other business it has its ups and downs. The second oldest football club in the country, it dates from 1863, and its Victoria Ground is in Boothen Old Road. They have produced more than thirty internationals, and Stoke is justly proud of Sir Stanley Matthews, for many the greatest footballer who ever lived. I am not sure that he is fully appreciated locally, but in a distant part of North Africa, the fact that I lived near Stoke, the home of Stanley Matthews, conferred almost a diplomatic immunity from the usual troubles that annoy tourists. The son of a Hanley barber, he epitomises the determination that is so much a part of the local character; there are still skinny kids who kick tennis balls

Hanley's Old Town Hall, its chateau-like facade shows its origins as
a hotel

The public market in Tontine Street, where potters once drew lots
for a chance to escape to the New World

Trinity exchange is dwarfed by the new telephone buildings

Unity House, Hanley,
Stoke-on-Trent's new
administrative building
emerges from the morn-
ing mist above the
remains of the past

Pretty cottages at the
entrance to Hanley
Forest Park, where
children play on the
massive spoil mound
of the Deep Pit

Stoke Town Hall, the most impressive public building in the
Potteries

Stoke's incredible public library and baths

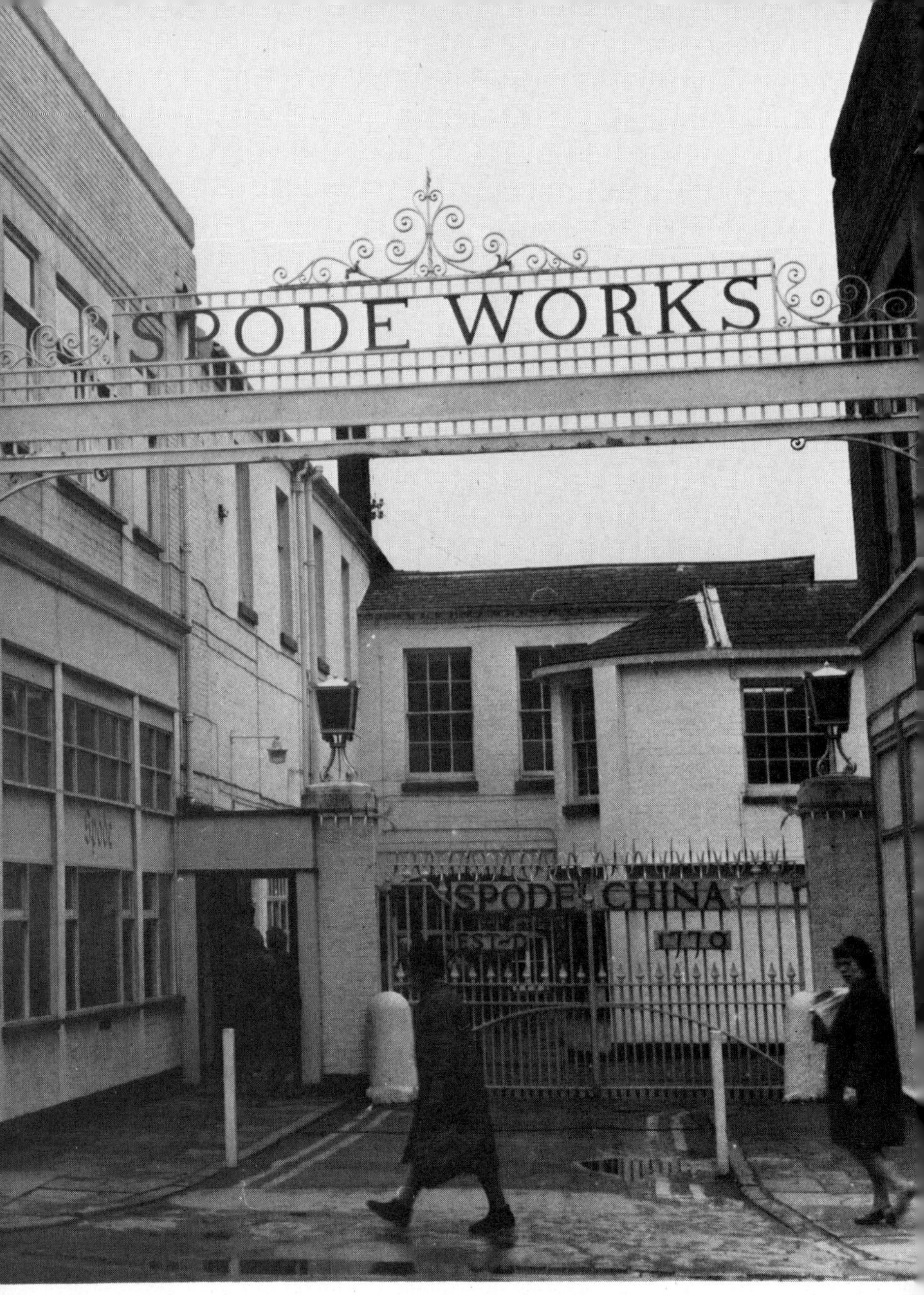

The Spode works which has stood in Stoke's busy main street for
over two hundred years

The 'Tin Man', Colin Melbourne's splendid tribute to 'the fight for Shelton Bar'

Josiah Wedgwood in Winton Square, turns his back on the North Stafford Hotel to face the railway station, the pride of 'knotty' architecture

Stoke Parish Church, the low, square pavement in front of the arches is Josiah Wedgwood's grave

The enigmatic Roundhouse, the sorry remains of Wedgwood's
Etruria

up and down back alleys, but I suspect that this is no longer the training ground of tomorrow's stars. None the less, the potteries' environment must still have something to offer: Stoke schoolboys won the English Schools' F.A. trophy in 1962 and 1963, and reached the final in 1971. Whether any of those boys will spend thirty-eight years at the top of the ladder remains to be seen. Perhaps one of them will take a vanload of trophies and records to Malta in retirement, and spend half the year coaching skinny, ragged Bantu boys in South Africa in the mysteries of the impossible body swerve.

From Boothen, the old Cut and the new roads lead to Stoke Station, possibly the most attractive railway station ever built, and certainly the best remaining in use. It was started in 1848 and it has not been messed about much since, externally at least.

Right at the height of the railway mania, in April 1845, the North Staffordshire Railway Company was formed. It was obvious from the initial surveys that the line would not only compete with the canal for trade, but would have to follow much the same course over the ground if costly tunnels and viaducts were to be kept to a minimum.

During the early days of the railways the canal companies had bitterly opposed the Acts of Parliament sought in order to construct the Iron Roads. They could well see that the speed, directness and carrying capacity of the new Wonder of the Age would seriously compete with what John Ward, in his 1838 defence of Brindley and his canals, called the "smooth and sulky passage of the barge, which one poor jaded horse laboriously tugs along".

Soon after the opening of the Grand Junction Railway, whose nearest station was at Whitmore Heath, the Trent and Mersey Navigation Company saw the value of their shares drop from £1,200 to £450, and dividends of 75 per cent shrink to nothing, maintenance absorbing the meagre profits. Where other Canal Companies merely faded away and died, the Trent and Mersey manœuvred most adroitly, both in dining-rooms and in Parliament, and managed to have themselves bought out by the new railway company for £1,170,000, thereby turning, themselves into a railway company, rather like the sorcerer's apprentice, who, after persistent practice, succeeded in lifting himself bodily by his own forelock, and transported himself out of trouble.

Another way of looking at it is the view that the Cut was absorbed by the Iron Horse in much the same way as the same apprentice sorcerer who hid in a bag, and then, to escape detection, pulled the bag in after him. The whole thing got off to a good start, the cutting of the first turf being made a gala occasion. The site chosen for this shindig was in Shelton New Road, not far from the present station, near the railway bridge in an area known as Winton's Wood. The railway bridge was not there then, of course; it was merely the last bit of moorland left in a valley that was becoming rapidly built-up.

The date was the last Wednesday in September 1846. All the workers were given a day off and expected to turn up and cheer. A respectable distance was maintained between the officiating party and the great unwashed by the erection of a rope barrier. The two Stoke M.P.s, J. L. Ricardo and W. T. Copeland of Spode's, were to perform together in the ring. Mr Ricardo was armed with a spade and Alderman Copeland with a wheelbarrow. The similarity to one of the popular prize-fighting contests could not possibly have escaped the delighted crowd, who obviously viewed the solemn officials as latter-day gladiators, Ricardo the Hastarius and Copeland the Carpentarius, to give battle to a poor sod in the arena.

All went as planned at first. The turf was deftly detached by four vicious thrusts of the intrepid Ricardo's spade and was being conveyed to the Aldermanic barrow for rescue when it was all too much for the crowd. They broke through the barriers and plunged into the fray, thumbs down, no doubt, intent upon tearing the luckless clod to pieces and carrying the bits off for souvenirs.

Alderman Copeland was wheeled off in his barrow by the ecstatic crowd; Mr Ricardo lost his hat in the confusion, but maintained his dignity and earned a great round of applause when he mounted his horse and rode from the ground with his spade over his shoulder like a victorious lancer's spear. After this there was a public breakfast in the new Town Hall at Stoke. All day the crowd celebrated with cakes and ale, amusing themselves with foot-races and gambling and fighting over the outcome of same. In the evening there was a ball at which popular numbers were played by the Duke of Sutherland's Quadrille Band, and the merrymaking ended with a brilliant display of fireworks.

Stoke Station was built in 1848, and was conceived as a part of a quite square where the carriages of the gentry could wait in peace for their masters, and came complete with an hotel, now the North Stafford, in which the weary travellers could be refreshed. The hotel was built in the style of an Elizabethan manor house with a Flemish flavour. The styles are rather mixed up, but the overall effect is very pleasant. The Station building echoes the mock-Elizabethan/Jacobean theme; it has a long, two storied Tudor-style front with gabled wings projecting from each end. The central porch has seven arches topped off with a square bay, and is decorated with Jacobean strapwork and cruciform shields. The windows have heavy stone mullions and, as one of my students observed, the pillars are Ironic. The square bay on the first floor housed the Board Room of the old North Staffordshire Railway Company, named the Knotty, from its emblem, the Staffordshire Knot. The panelling and ceiling are wonders in an area rich in such treasures. The opulence and precision of these plaster mouldings is not hard to understand. The pottery industry trains and employs the finest plaster-of-Paris workers in the world. Appropriate, therefore, that local architectural plasterwork is second to none. The lovely, but atavistic, mock-Elizabethan/Jacobean design was used on all the Knotty's main stations, and are well preserved at Stone and Cheddleton.

In the middle of Winton Square, which is where the station is, is a bronze statue of Josiah Wedgwood. Characteristically, and true to his non-conformist principles, he has his elegant, frock-coated back to the abode of Satan in the licensed premises behind and boldly faces the progress enshrined in the new railway which, at this point, not only overshadowed but overlooked his canal.

Josiah is shewn with two good legs so either it is a portrait of the artist when young, or he had a very good turner. Still, it remains a matter of opinion.

The unveiling of the effigy was another grand affair. The statue was erected by public subscription, and was sculpted by a Mr Davis, of London. Once again a public holiday was declared, a guard of honour of the county regiments and volunteers was formed, supported by the local militia and a battery of the Royal Artillery. The civic heads of all the local towns were invited to sit on a platform that had been erected in front of the station. The Duke of Sutherland was invited to perform the ceremony, and anybody who was anybody, or thought they were, was crowded on to the balcony of the station, over the main entrance. At the moment of unveiling, a salute was fired, whereupon the horses in the nearby railway goods yard bolted, causing considerable damage, a barge was towed at a great rate of knots by its stampeding tow-horse until it rammed the lock-gates higher up the canal, and Winton Square and all its occupants were choked with gun-powder smoke which lasted about ten minutes as a dense fog. When the smoke had cleared several gentlemen in the crowd had lost their watches and handkerchiefs and Wedgwood was wearing a floral bonnet.

The volleys of musketry and the cannonade caused a stir across the Trent in neighbouring Fenton. As part of the growth of the railway services at Stoke Station, an enormous stable had been erected which, according to the *Staffordshire Advertiser* of 22nd April 1848, was the largest in the country. This was only part of the vast range of buildings associated with the railway. It has all been destroyed.

IV

FENTON

The name, in Anglo-Saxon, meant a settlement by a marsh.
As you go by Stoke Church eastwards and pass Whieldon
Road, you are going into the township of Fenton Vivian. A
little further, under the railway bridge, and in a quarter of a
mile, over a railway bridge, and it is now Fenton Culvert. The
ground is quite low-lying and it is easy to cast the mind back
a thousand years or so and understand the bit about the fen
or marsh.

With such charming names for some of its parts, it is
difficult to know why it was forgotten. There are two principal reasons. One is that the main road runs right through it,
whereas in the other five towns it wanders about in such a
way that the town impresses itself upon you, often because
you recognise places you have recently passed in an attempt to
get out. In Fenton, Stoke is left behind and Longton has
sprung up before Fenton seems to have been.

The other reason is Arnold Bennett. Whilst it is easy to
enjoy his works as masterpieces of English Literature written
by a story-teller of uncommonly compelling powers, it is
difficult to accept that he was really writing about the
Potteries people, or that he even liked them. Describing a
crowd of Potteries folk in *Clayhanger* he said they were

a shabby lot; persons without any social standing: unkempt idlers,
good-for-nothings, wastrels, clay-whitened pot-girls who had
to work even on that day, (a holiday) and who had run out
for a few moments in their flannel aprons to stare, and a few
score ragamuffins, whose parents were too poor or too careless
to make them superficially presentable enough to figure in a
procession.

I suspect that there was nothing in 1910, when *Clayhanger* was published, to recommend tired-out, honest, useful old Fenton to the comfortable middle-class memories of a Burslemite who could not get away fast enough from the persons without any social standing and the clay-whitened pot-girls.

To lovers of pottery, Fenton must be almost a place of pilgrimage, since it was here that Thomas Whieldon, the central figure in the eighteenth-century drama of ceramics, had his home and works, and where he taught so many of those who went on to develop the industry, enriched by his many-sided genius and open-handed generosity.

He was born in 1719 and by 1740 he was a master potter with works at Fenton Low. Fifteen years later he had expanded so much that he had to have a further pot-works at Fenton Hall. The output of his factories was amazingly diverse. Excavations on the site have shown that, at one time or another, nearly all of the varied kinds of pottery of the eighteenth century were being made there. This does not mean that they were all being made by Whieldon. In common with his contemporaries, he often let out bits of his works to other potters, who may have been making ware he was not concerned with.

Many of the types are found on his sites in an earlier context than anywhere else, and he may have originated some of the varieties of earthenware that were to become the staple products of the industry. About the only things we can say with any confidence that he did not have a go at were transfer-printed ware and coloured enamels, although he lived on, potting away energetically, until long after both processes had become well established; he stuck to what he knew.

He was obviously considered by potters of his time as the greatest potting genius alive, and aspiring young potters thronged to his works to benefit from his abilities as a tutor and also so that they could use the fact of their employment there as a recommendation in seeking advancement.

The most famous of his apprentices was Josiah Spode who was indentured on 9th April 1749. Thomas promised to give him an extra threepence a week, if he deserved it. Spode's success was due in no small part to the thorough and benevolent tuition he received in Fenton.

The Queen's Ware that made Wedgwood's fortune and reputation was developed by Whieldon, who favoured a decoration of random coloured glazes, where Wedgwood liked the effect of the brilliant, yellowish lead glaze. Whieldon used that as well, of course. He employed and trained several men who were later to become famous as model makers, among them William Greatbach and his father, Daniel. Aaron Wood of the Burslem family learned his art with Whieldon and, with the Greatbaches, went on to spread the fine tradition of figure-work developed at Fenton Hall from the delightful little military figures made in salt-glaze and slip-ware, in which that other elusive and shadowy Fenton genius, Thomas Astbury, seems to have had a hand. Doulton's figures have a long and respectable pedigree.

Considering the number of Wedgwoods there were who were concerned with the industry, there must have been something very special going on in Whieldon's works to persuade Josiah to make the journey from Burslem to the other end of the potting world to seek a partnership. The association would not have been one-sided. Those whom Thomas took under his wing were usually destined for greatness or success, or both. The youngest son of an enormous potting family like the Wedgwoods would be expected to have something to offer, and the wily Whieldon must have had his talent scouts out. The Fenton firm were well to the fore in the development of clay additives, like flint, and were known to be consistently successful in their standard of finish. Wedgwood stood to gain a thorough knowledge of the most modern techniques and processes, which were certainly not being used by his relatives, with whom he could, presumably, have sought his first partnership, as he had his apprenticeship.

Whieldon was aware of the high degree of skill the young Josiah had developed, and also was bound to benefit from a proprietory interest in the many experiments he was making. Experiments that could be properly pursued in the extensive, well-equipped workshops at Fenton Hall, which were sufficiently affluent to enable the budding genius to work unfettered by considerations of cost.

When we remember the large number of potters of influence and taste who were trained by Whieldon, it is fitting that part of the site of his works is now occupied by the Stoke-on-Trent Sixth Form College. If geographical locations can have an aura, that of the Manor of Fenton Vivian is surely one of education and progress.

The magnificent provision made for the 'A' level students at the College is a far cry from the educational provision made for the children of Fenton over the years.

Fenton remained an area of fields and scattered pot-works until late in the nineteenth century; in fact, there is still much open ground, some of it the remains of farms, some of it the spoil heaps of coal and iron working.

The first school in Fenton was built in 1839, much later than most other towns, and reflects the sparse population rather than any lack of concern. The school was Fenton National School, originally built in the churchyard and designed to cater for Boys, Girls and Infants. At that time education was not compulsory, and many who should have known better doubted if it was desirable for the bulk of the population as it was supposed to give them ideas above their station. The good folk who ran the nonconformist chapels did their best to encourage children to go to their Sunday schools, where an attempt at a basic education went hand in hand with Bible study. But since children from the age of five often worked a sixty hour week and walked miles to and from work, their Sunday 'lie-in' was an absolute physical necessity.

The extent to which education was a concern amongst all ages in the 1840s is shown by the number of times the answers in the Report on Child Labour of 1842 refer to schooling. John Orton, aged eight, who worked for F. and R. Pratt and Co., of Fenton, said that he could neither read nor write, once went to church school, but stopped going because he had no 'trowsers'.

The next thing that happened, educationally, presumably for boys who had 'trowsers', was the opening of a branch school of the Stoke St Peter's Church of England School. It was opened in a carpenter's shop in Bridge Street in 1858, then it was moved to a house that had been used for religious meetings by dissenters. All this was in Mount Pleasant, an area of the ancient Manor of Fenton Culvert, more recently known as Great Fenton, in the region of Smithpool Road. Then, in 1861, on the site of the present Mount Pleasant C.E. Junior and Infants School, an iron church-cum-school, named St Paul's, was erected. In 1867 the first part of the present church and school was built.

Up to this time not much had been done to alleviate the misery of children working in the area, but some of the disadvantages of infant slavery were becoming obvious to a few manufacturers with the ability to think.

In 1862, twenty-six manufacturers signed a *Memorial of Employers in the Potteries*, and sent it to Sir George Grey, the Home Secretary. They said that they wanted him to know that children were being employed in the pot-banks at a very early age, which was injurious to their education.

We learn from this memo that most of the children were taken away from school before they were ten, and that in fifteen out of the twenty-one National Schools in the district the average age of Class One was ten years, two months, and four fifths of them left before they reached the first class.

It is almost uncanny to see the workings of a clear, statistical mind at such a distance. They looked at 116 pot works, and found that in 22 of them there were 177 children under the age of 10, and 576 under 13. They concluded from this that it was the employment in the potteries which caused the early removal of children from school, not the 'school-pence' that had to be paid, nor the cost of books and, presumably 'trowsers' for boys and unmentionables for girls.

In all of this, of course, the parents were not breaking any law, since education was neither compulsory nor free.

Our memorialists went on to show that there was a "vast amount of ignorance", in that out of 670 working children questioned, 185 could neither read nor write. They concluded that the employment of children at so tender an age was injurious to their health, stunted their growth, and caused in many a tendency to tuberculosis and distortion of the spine, as they had "the evidence of competent medical men to testify".

They deplored the evils they had mentioned, and regretted that it would be impossible to get the manufacturers to agree to right the wrongs without legislation. They urged the Honourable Principal Secretary to some "legislative enactment" to prevent the employment of children at such an early age, and to secure for them at any rate a minimum of education. Amongst the signatories were Mintons and Wedgwoods, but not Spode, and none from Fenton.

This was the mood of many throughout the country, and the agitation caused by much good-hearted concern led to the Forster Education Act of 1870. This introduced compulsory education and saw the founding of a state system of elementary education. And elementary was the right word for the education they got. Children were taught to read, write and do sums. They also had to provide their own materials and pay for the privilege. With the constant epidemics caused by bad sanitation and sheer pig-ignorance, not many people could actually afford to send their children to school, even when the parents had work, and employment was a variable commodity in the 1870s and 1880s.

Following the Act a number of new schools were set up; many of the buildings exist today, in one form or another, and they usually tell their own tale. During this period the parts of Fenton that are now built-up were being built, and the empty spaces along the roads were being filled with potteries. Because of this, the bulk of the town that grew up out of the coalescence of the old villages and hamlets was hidden from main road travellers, and it was possible to overlook the fact that an urban centre even existed. Not that the urban part of Fenton was very attractive; because of its position between Stoke and Longton and its low-lying characteristics, much of it was an open sewer for other people's overflow of effluent. In 1839 Fenton, together with Stoke, Longton and Trentham, had a body of Improvement Commissioners appointed whose job it was to see to the policing and lighting of the area, and to set about generally improving the streets. By 1850 they had not done much to help. One of their sub-committees, the General Board of Health, had a good look-round and found that the area abounded in ash-pits, communal privies—some were six-seaters—and open sewers which were really random rivulets of filth that ran in front of the houses, many of which had open privies directly in front of their front doors, which opened into their only room. There was, of course, no piped water and amongst all this insanitary squalor, water was obtained from wells. It was not until 1849 that her neighbours allowed Fenton a grudging pipe of water, and that was only to public taps, it reached hardly any houses; little wonder those who could afford it built their houses on any bump of high ground they could secure. So, in 1870, Fenton was like half of a broken tea-saucer, with the Trent running along the broken edge, and Fenton Low, Botteslow, the Foley, Mole Cop and Mount Pleasant as the outer rim, decorated with large houses. By 1853 they had a few sewers laid in China Street, and within a couple of years they had pushed their sewers into Park Street, King Street and High Street, which is in the old Lane Delph area of the town. Despite the claims of a number of misguided etymologists the word 'delph' has nothing to do with Dutch pottery, but is derived from the nice old English word 'delve', which meant to dig, and from the 1550s onward someone was digging something in the area. At first ironstone, then clay and coal were dug—or delved—all over the place, right up to the mid-twentieth century. In fact, Fenton was at least as important for its ironstone and coal as for its pottery over the past two hundred years and the dereliction caused by the mining and quarrying is responsible for the town having so many large patches of open ground. These patches are like wild, enormous tennis courts of red ash, supporting beautiful seasonal displays of thistles and rose-bay willow herb, and, surprisingly rectangular plots of common blue lupins. These lovely naturalised flowers take over abandoned allotment gardens, and show us where the miners and potters, inveterate gardeners and pigeon fanciers, tried to wrest something of use and beauty from the lunar landscape they helped to create.

This sort of scenery can still [1976] be seen on the left

behind the wall that emerges from the railway bridge on the main road out of Stoke into Fenton. Set into the wall are the original gate pillars of the drive to Fenton Manor. Opposite, on the right, is a huge new trading estate, built into the bow formed by the road and the railway line. These large new buildings are excellent space-gobblers and conceal the dereliction left by the destruction of Fenton's great railway buildings and sidings.

Moving up into the town, again on the left are the local Workshops for the Blind, opened in 1934 on the site of the old Rialto Works, which had been the premises of the Pratts of pot-lid fame. Felix Pratt, of Fenton, introduced a new form of decoration in 1790. This consisted of designs painted in green, yellow, blue and brown, using metallic oxides that could be fired at high temperatures in a glaze in which they did not dissolve, which was the problem with the kind of colour used by Whieldon and others.

Pratt usually had his colours painted on to decorative plaques and jugs moulded in high relief. The motifs are frequently fragments copied from famous prints of the period, or caricatures of people in the news and nursery rhyme characters, which is probably much the same thing. Favourite subjects were Humpty-Dumpty (the Prince Regent, later George IV) and Nelson and Wellington. Gainsborough Ladies and Little Miss Muffett were popular.

The family Pratt had not done with colours yet, though. To get a full-colour effect in the early part of the nineteenth century, potters either had to have the pieces painted by hand entirely, which was very expensive and relatively slow, or they had the pattern printed in outline, usually in underglaze blue, and then filled in with colours by hand, after the manner of a child's painting book. They were not tied to blue, although it was the most popular, but by about 1830, prints were possible in black, which was and still is rather tricky, and various shades of yellow, green, red, brown and purple. Combinations of these were tried by many potters, but it was the firm of F. and R. Pratt of Fenton who perfected the first really successful method of multi-colour or polychrome printing, about 1840. They soon had a wide range of popular patterns designed by Jesse Austin, and the pot-lids and tablewares that resulted from the association are now eagerly sought after by collectors all over the world.

Very much a part of the character of Fenton is the Royal Oak Inn. It has stood at the crossroads of Christchurch Street and City Road, with the evocative Manor Road opposite, since before 1818. This is another occasion when it pays to look up. The proportions are perfect. Above the Victorian pub windows, which are not too bad, are three graceful Venetian windows under a pedimented gable. The stucco of the ground floor will probably change colour but the bricks of the upper part have been cleaned and are that clear blushing red that is typical of the best potteries bricks. A recent commentator has said that the Potteries are built of ugly blue bricks. He is either blind, lying or malicious. There are fine, glossy purply-blue bricks used either for the whole building or for decoration, but the overall impression of Potteries brick work is of warm, welcoming red bricks, generally very well laid.

On the opposite corner from the Royal Oak is the building that used to be the Athenaeum, built in 1853 to the design of a local firm of architects, Ward and Sons of Hanley. The lower part of this sturdy rectangular building is of stone, done in the Italian manner, and the top is of warm brick with the obligatory stone dressings. The front, on Christchurch Street, has five bays and the front door, in the middle, was borrowed from a Doric temple.

An athenaeum was a literary or scientific club. Judging by the size and opulence of the building, there must have been a big demand locally for learned discussion, at least at first. The Club seems to have thrived until 1867, when at least part of the premises was being used as a school by the older boys from Fenton National School. Sometime around 1868 the building was acquired by William Meath Baker, one of the Potteries' great men. It is not known to what use he put the Athenaeum between acquiring it and 1873, but in that year he agreed to the new Board of Health taking it over as their offices. In 1888 the Board of Health moved down the road to the new Town Hall, and Mr Baker converted the building for use as an Art School. It continued to serve in this way for many years until it was acquired by the District Bank. It is now [1976] one of the most attractive of the premises of the National Westminster Bank, who have some of the finest buildings in North Staffordshire.

Just along Christchurch Street is Albert Square and on the left of Albert Square is Fenton's charming Town Hall. Whenever I see it I get the impression that it was built back-to-front. Whether or not this is so, it was erected in 1888. It is a brick building with two stories and, of course, stone dressings. The design is a remarkably effective compromise between the styles of domestic architecture of the time of Charlemagne and Charles I of England. The Hall itself has six bays with a central gable which has gabled wings flying away at each side. In the upper stories there are minstrel's windows that a respectable guide book would call oriels.

The Town Hall was built by William Meath Baker of Athenaeum fame. His first tenants were the Board of Health, and then the Urban District Council moved in, also as his tenants. In 1897 the Council bought the building from the obliging Mr Baker, whose generosity and public-spiritedness knowing no bounds, proceeded to give the land for a library in 1906. The library, which was free, was built at the expense of £5,300 by Andrew Carnegie. They named the street in which library is to be found, Baker Street, which same thoroughfare was appropriately enough chosen for the front door of the Police Station when it, too, was tacked on the back of the town hall in 1914. It is to this building that naughty local motorists furtively creep to deposit their pleas of guilty, and the same to which they steal at the dead of night to deposit their little bundles of pound notes when justice is finally done.

It was in answer to an enquiry about this particular building that I first heard it called a 'tine owe', just about my first brush with the Potteries dialect. When you ask for directions in the street, you will find people eager to be helpful. How profitable this is depends to a large extent on the age of your guide. This is where dialect usually takes over. The middle-aged, with a diet of wartime news broadcasts from Frank Phillips, Bruce Belfrage and all, at a formative

stage can manage 'fer tow crate' meaning to speak cor-
rectly—for talk right; they usually use 'for' in place of 'to'. But
if your helper is old or young you may still be lost at the end
of half an hour's hard help—and they do take the trouble.

Potterese is as much a way of life as a dialect to the older
ones and is certainly not intended to confuse. The sharpening
of the vowels and a sprinkling of 'thees' is more Northern
than Midland, and hints at an isolation only recently relieved.
The cultivation of such a tongue amongst the young is at once
a deference to their elders—the fifth commandment is upheld
in the Towns—and the confirmation of an identity that defies
the late loosening of the bonds of the moors.

When they are doing their best, or worst, depending on
your point of view, 'council house' sounds much more like
'kind slice'. Try it. 'Kind slice'. Good. Spoken like a native.
Anyone interested in our language will be fascinated by the
survival of Anglo-Saxon words in 'Potterese', though not all
are polite.

There is, fortunately, a good 'Primer' in Potterese in *Arfur
Tow Crate in Staffycher* (How For Talk Right in
Staffordshire), by Andy Ridler and Alan Povey.

Returning to the main road and moving towards Longton
brings Fountain Square into view. It is the next crossroads
and it certainly is not square. There is, however, an oddly
shaped, paved area on the left with some substantial mid-
nineteenth-century houses of the sort that would have been
rented to the better paid workers of the factory opposite.
Standing on those blue dust-bricks, looking over the road
towards the factory, one has a real impression of the
Potteries of one hundred and fifty years ago. There must,
at one time, have been dozens of squares just like it, all
over the Potteries. Leading into the square are some low,
early nineteenth-century cottages that have now been converted into shops, but looking up, the feeling of age and
distance is easy to acquire. The factory itself is small, as most
pot-works were. It has only three bays, the central one the
'tunnel' surmounted by a Venetian window under a gable.
This ubiquitous potteries window is not merely a fashionable affectation, it is very sensible. For one thing, it lets in

plenty of light which is a good thing for a window. Then, its sashed frame is easy to remove, and the wide open space that removal leaves is ideal for the reception of large articles like machines that need to be replaced from time to time, and which could not possibly have been taken up the rickety steps and along the mean, narrow passages that are a feature of even quite modern pot-banks. Remarkably heavy loads could be hoisted into those windows on a sturdy, makeshift gantry which was easily supported by the strong arch beneath. And added to all this gratuitous utility is the benefit of the most dignified and beautiful style of industrial architecture anywhere. The fact that it is not original, that it combines both Classical and Gothic elements, does not rob it of its charm, even though the blush of the bricks does sometimes make it seem a little self-conscious.

Speaking of makeshift gantries reminds me of the striking gift of mechanical improvisation that the potters have, a gift they share with their indispensable colleagues, the miners. Although this 'Heath Robinson' nature is seen to a minor extent in the allotments and gardens of the Towns, in the quaint sheds and pigeon-lofts with their homemade fastenings, hinges and plumbing, it is only at work that the full extent of their inventive ability is appreciated. It probably stems from the need of potters, especially modellers and flat-ware pressers, to make their own tools to 'profile' and shape their work. Those who had developed the necessary skill or touch would take home a thin piece of lightly fired ware and, working for hours with files and stones, shape the piece into a silhouette of the back of a plate or saucer. If it worked he would be rewarded with a few pennies by the works, or sometimes they were sold to neighbours, potters from other banks who were glad to pay a little for the extra productive capacity that a good new former would give them. Usually, after hours of painstaking, meticulous work the slender fragment would break. Still, as one old lad told me, it had kept him out of mischief. Slate was sometimes used for this job, and so was iron, but both tended to mark or stain the pot, whilst no harm could come from the minute amount of clean 'body' that would be left by a former made from clay. As well as this obvious adjunct to potting there are the one-off contraptions knocked-up to make life easier or a job quicker and more efficient. Tool rests, water drippers, splash guards, lamp shades or reflectors all have their little part to play; most are fashioned from scrap materials and their number and variety is infinite.

With all this mechanical ability, and it has been manifest for hundreds of years, it is hard to understand why there was such a resistance to machines when they were first introduced. After all, it is certain that the machines themselves could only have been invented by a working potter, whatever his status in the hierarchy of the firm.

Eastward from the Square is Victoria Place, where City Road and Victoria Road unite to become King Street. There, on the corner opposite the public conveniences was Mason's Ironstone Works, in the old township of Lane Delph. As with other firms who made their mark there was not just one Mason; there were several.

Miles Mason came to the Potteries from Liverpool via London about 1800. After about five years he had built up both a good business and a high reputation for the quality of his blue-printed china. He was a very enterprising gentleman, having another pottery in Liverpool and a shop and warehouse in London.

In 1805, he took over the Minerva Works a little further up the road towards Longton, a fine building that has been the home of many first-class china firms, including Pratts, and, most recently, Crown Staffordshire China. However, back to the Masons. Miles was the father of George Miles Mason, Charles James Mason and William Mason. Old father Mason—Miles—seems to have been an unqualified success with almost the Midas Touch, and he kept out of trouble of any sort. In June 1813, Miles was succeeded by George M. and Charles J.; this latter had been working on a new 'body' for china at one of the several works to which the firm had spread, Daisy Bank. Having made his father retire, Charles J. took out a patent for "a process for the improvement of the manufacture of English porcelain, using scoria or slag of ironstone pounded and ground in water, in certain proportions, with flint, Cornwall stone, and clay, and blue oxide cobalt". So Ironstone China was born. They took up more premises in Broad Street, Hanley, and seemed set for as long a run of success as their father. The Masons of Lane Delph, together with their partner Sam B. Faraday, founded the Fenton Mechanics' Institute in 1839, and attracted a good audience of serious-minded workmen; so they seem to have had the interests of their employees at heart, the tame ones, anyway. By 1840 G.M. had withdrawn from the partnership, and C.J. was left to soldier on on his own. He was singled out for the full fury of the attack of the Chartists in 1842. His house at Heron Cross was completely destroyed by fire. Our old friend Evans, the crusading newspaper-man, had singled C.J. out as one of the worst employers in the district. This was on two counts. First, he worked the allowance system; second, he had introduced a fiendish machine for pressing flat-ware, the jolley. He had also been the chairman of the Master Potters' Organisation during the catastrophic strike of 1834. Added to this, he had married Sarah Spode, whose family owned mines from which the colliers were locked out in 1842. He had more than a little interest in local mines himself. So, he had earned the animosity of both large factions of workers and he suffered accordingly.

On top of the heavy burden of loss of his personal property, the machine that had caused so much trouble proved to be unreliable and had to be withdrawn, a total loss. Short of capital and hated by his workers, C. J. Mason became bankrupt in 1848. Such was the quality of his work, and his father's before him, his successors chose to retain his name on their wares, and the Mason trade mark continues to this day.

As for the potters and their view of machines, according to the *Potteries Examiner* of 1844, because of machines cellars were crowded with six or eight living human skeletons; naked people were exposed to frosty winter winds; hungry children cried; hearth-stones were desolate and cupboards were empty.

The Potters' Central Committee issued a poster calling for a fund of *five thousand pounds*. Machinery was the potter's deadliest enemy. Obstructing its introduction into any branch of the trade should be the greatest principle of life. Five thousand pounds sterling was required from the seven thousand adult potters of the district. Mr Mason's hands would legally leave their work if they were supported, eight half-crown levies would do the trick.

They did not raise the money, and Mr Mason's hands stayed at work. Not that it helped Mr Mason much.

The problem is easy to understand. One of the early machines for making plates was guaranteed to succeed as far as manufacturing was concerned. When it had been in operation for two or three months it was making very good plates. In three or four days it made six hundred dozen, an oven load. Every time the machine was altered it improved, and was soon able to make plates at the rate of six hundred dozen a day. Or so the agitated workmen of Leeds believed, and they communicated their agitation and belief to the Staffordshire Potteries.

Potting is a complicated process that, on an industrial scale, involves far more than clay. The decoration often requires the skill of printers as well, and large fortunes have been made not only by potters, but by paper-makers, printers and colour-merchants.

A printing machine installed at Messrs Mellor, Venables and Co., Burslem, had put ten adult printers out of work and made work for twenty-two transferers. There was an indication that the same firm had it in mind to introduce an even bigger machine that would make work for forty women transferers. This was all too much for the ladies concerned, who were not able to take a moment's rest now that they were served by the blacksmith's monster. They went on strike for easier times.

As a result of the gradual invention and introduction of machines, the pottery industry changed but the labour force grew. Not only that, a large ancillary engineering industry grew up, providing work for more people, and creating an even greater demand for iron and coal. These two were available in good quality and large quantity in Fenton, and many colliery owners were also iron-masters, smelting at the pit-head the ore with the coal that came up the same shaft.

One of the biggest landowners in the area was the Broade

family, who had been prominent since 1564 when a John Broade purchased the estate of Fenton Vivian from Thomas Essex. It was from the Broades that Whieldon purchased his land; the other landowners of consequence in the area were the Fenton family, whose variously-named descendants exploited the Fenton Park estate next door to the Broadfield estate of the Broades.

Coal had been worked at Lane Delph from very early times; this was another area where coal outcropped, or came to the surface. In this case it was a good quality coal from the Knowles seam which came to the surface somewhere between the present cemetery and Pool Dole.

By 1864, eight seams of coal had been worked on Broade's estate, six had been worked out, one was not worth getting; one, the Deep Mine seam consisting of both coal and ironstone was "but little worked out" and fifty-five acres were left to get.

The names of the seams are quite picturesque. In the Broadfield colliery they are: the Bassey Mine; the Peacock; the Spencroft; the Great Row; the Cannel Row; the Deep Mine; the Knowles and the Ash Coal. Under these, and not being worked in that area in 1864, were the Mossfield, Yard Coal and Birches. Most of these seams were named either from the pit where it was first won, or where it was most significant. Some were named for a peculiarity of the seam or a special quality of the coal.

In a report dated 7th June 1864, from W. S. Cope of Hallfield Cottage, Hanley, to F. S. Broade, Esq., of Fenton Hall, a complete analysis of the coal on his estate is given. This was on the expiration of the lease of Copeland and Baker. Mr Cope said that the Peacock had not been got at all, in consequence of it being bad, thin coal, and a heavy roof.

The conclusion of this report raises an interesting point. It deals with the engines and machinery in place in the mines. He says that if the pumping engine at the Cannel Mine Pit and the winding engine at the Knowles had been anything like adequate for making the recovery and working out the same, it would have been his duty to have recommended that they be "taken at a valuation". But, on the contrary, he found the pumping engine in bad condition, in fact worn out, and the pumps had left their stays and were all gone to the bottom of the shaft and were under water, and most likely the pit sides had gone down with them.

That is the point I find most interesting. As we go up Victoria Road towards Longton and pass Fenton Park, over on the left is where the pits were. No one knows just exactly where the shafts were. Under the ground are the tunnels in which men and women, girls and boys crawled on their hands and knees to get the coal, and down there with their ghosts and the eternal echoes of their agonies are the pumps that eventually failed to keep the water out and drowned the pits. And down is the right word. Mr Cope found the top of the water at 978 feet below ground level. He failed to speculate on how deep the water was.

I love his description of the winding engine at the Knowles pit. He said he found it to be one of the "oldest of the old", being an old open top cylinder and not of the least use for further working of the pits. It is a fascinating thought that so long ago mines had machinery that was a hundred years out of date, and that it could be allowed to fall down a shaft, taking the shaft in after, and be permitted to stay there.

Most mining areas claim to have the deepest pit and North Staffordshire claims nothing less in that the Sutherland Pit, in Fenton, was most likely to have been the deepest pit in the past, since it was 3,318 feet deep. I say in the past because there are even deeper seams below those yet to be worked. Modern methods now permit the further exploitation of seams that could be only partly won in the past, some by deep mining, some by opencast mining, a process in which the top layers were cut away to reveal the coal underneath. In the stripping of the 'overburden' many small shafts and burrows come to light showing how busy were our ancestors, and we see how little they actually got for all their hard work. Not only do their old workings appear, their tools and transport, baskets and tubs, are revealed, and sometimes, sadly, their bones. Today's opencast miners with their powerful tools can work to within a fraction of an inch, recovering coal too thin to work by older methods. Although there is a lot of dust and noise whilst they are at work, they are quick, and since the opencast is almost always in an area of old workings, when they put all the waste back, they put in proper drainage and smooth the surface, replacing a lunar landscape with an area of park. I asked a driver of one of the big earth-movers if he was not worried about falling into the old workings. He answered nonchalantly that the galleries had usually collapsed anyway, and that "the old men" never dug a shaft wide enough to take his machine, especially side-ways.

In an area that has been so intensively exploited for its minerals for so long, and in an industry that not only rarely kept accurate records but frequently covered its traces, the occasional surprise discovery of old workings is not uncommon. One of the largest of the Fenton Pits, the Stafford Colliery and Ironworks at Great Fenton, worked seams that had previously had small-scale pits and a collier there described breaking in to the old 'roads'. They knew that they were likely to come across the "old men's work", but not where, nor when. Since it was a dry seam they were not worried about flooding, but knew there was likely to be gas, and probably foul-air. When they broke through they were making a gallery between two of the modern roads when they stepped into an old 'stone-drift', which had produced iron ore at some time. The floor was at the same level as theirs, but the roof was much lower. The air, he said, was quite good, and there was no trace of gas. The absence of deadly, highly explosive methane would have been no surprise in a stone-drift, since the fire-damp is often produced by the decomposition of loosened coal and is more often met with in the coal seams. A stone seam could have produced gas, however, since there were often thin seams of coal in the stone, and that would be all that was needed.

There was no trace of the previous tenants, except their tool-marks on the tunnel sides, but despite this and the restricted nature of the place, he said that they hesitated to go into the drift, not out of fear or trepidation, but from some feeling of respect, as though they were on hallowed ground. Almost like going into a strange church, or on to a clean new carpet with muddy boots. The old tunnel had to be sealed off, partly because it would interfere with the ventilation of the new workings, partly to prevent anyone straying down it and getting lost. Most pits have their stories of men straying into old workings and vanishing without trace. Although there may be no smoke without fire, I have yet to be shown proof of this ever having happened, except in the rich imagination of the colliers, always ready for a good yarn, especially if it has a grisly ending. The surprising thing about the story from Fenton's Stafford Colliery is that it is not thronged with skulls and bones, and last despairing messages to new brides scratched on long lost shot-boxes. This colliery, like Fenton's other giant, the Glebe, has now ceased production; the seams that were worked from them are now being won by other pits further out. With the disappearance of the pits it is all too easy to forget that North Staffordshire is as much concerned with coal as ceramics. And, although there is no longer any sign of it above ground, men are still toiling away far below, miles from the hole into which they descended.

Going down Grove Road, to where it becomes Duke Street at Heron Cross, and then back to the main road again, we are on the old Roman road, Ryknield Street, which goes on through Longton to the Roman Fort at *Derventio*, our Derby.

The last stop in Fenton is the Foley. How it got its unusual name is uncertain. There were some Lords Foley, who may have been named for the place, or who may have had the place named after them. It has been suggested that the title might have something to do with an eccentricity of some former style of architecture, but it would be folly to speculate.

There are three superb pottery buildings there, and they are not too conveniently named. They are: the Foley Potteries, the Foley Pottery and the Old Foley Pottery. To unscramble which is which, as we approach the border with Longton, the Foley Potteries are first, on the left, the Foley Pottery is next, on the right, and then, also on the right, is the Old Foley Pottery, although they are all old, and are all at the Foley.

The Foley Potteries, the big one on the left, were built by John Smith about 1820. They were most up-to-date, being described by a contemporary source as a "modern and very complete works, combining a Steam-engine, a flint-mill, and every other convenient adaptation to the business". The 1842 commissioners found the buildings to be "modern, well constructed, open, roomy and in all respects good". What we see there today looks very like the standard Georgian-Regency pot-works, except that it is so long. There are two Venetian windows, with pedimented gables, and each has a large elliptical arch underneath. The building seems to be all of the same period, but the western half, the first part, was built some fifteen years or so after the original eastern half. It seems that the building Mr Smith put up was not adequate for the purposes of the Wilemans, who then owned it. Now, it is big enough to house a china wholesaler, the firm of China and Earthenware Millers Ltd, who run the mill, a tile making company, an earthenware warehouse, and an upholsterer's and cabinet-maker's factory.

Next is the Foley Pottery, on the right, built about 1790 by Josiah Spode for Samuel, his eldest son. This was the one who sent his men to gaol over a dinner-hour dispute, and who made a public spectacle of himself by dancing a jig on his workmen's clothes, no doubt in the road we still use. The building is the standard one with its archway, Venetian window and pedimented gable, and is now the premises of a wholesale and manufacturing chemist.

On the same side is Foley Place, built in the 1830s, and not meant for the poorer workers. It is an L-shaped block of two-storied houses, finished in stucco. The houses are Georgian in style and once had a communal garden, which is now a garage; there is a pub, the Foley Arms.

Finishing off this tantalisingly incomplete little corner of Georgian England is the Old Foley Pottery. It is the right shape and size and must be approaching two hundred years old. It was in the yard of this works that the good man, John Wesley, preached one of his last sermons on Sunday, 28th March 1790. The bank was then in the possession of Mr Myatt, and passed, as such an old building must, through the hands of many manufacturers, who, like most local potters, have maintained an astonishingly high standard of production, decoration and design. The most recent occupants, James Kent, have been there since before 1900, and had connections going back many decades into the mid-nineteenth century. This is another instance of a firm that does not have the widely known name of a small handful of local potters, but whose work has been consistently good, and who have worked on an historic, but difficult, old site for more than three generations. It is the smaller, little known firms that make up the real character of these old towns; they are the massive bulk of the pyramid that enables the few at the very top to remain there. It is because of the good, solid work that goes on generally that the superb, breathtakingly lovely best china can be produced. It is also true that many respected old firms, with their special gifts and styles, have been taken under the wing of the larger manufacturers, each contributing something extra, each usually able to pursue its own individual character, and each adding another facet to a ceramic jewel.

No other person has had so great an influence on the character of the Potteries as John Wesley. When all the workers were trooping into the Fowlea Valley in the mid-eighteenth century, there were only the two parish churches of Stoke and Burslem, and two private chapels, one in Hanley and one at Lane End, to cater for the spiritual needs and general welfare of the immigrant and native populations. Even the most entrenched and reactionary of commentators admits that the Anglican clergy of the time was neither vigilant nor pious, as we have come to expect our priests to be. It says much for the provocational ability of Wesley that the Established Church pulled itself together and built so many churches and schools in the first half of the nineteenth century. But before this happened, a rash of chapel building had broken out all over the place.

Many of the buildings remain, in altered form; many have long since served their purpose and ceased to exist. However, the paths beaten to their doors have become roads, and the civilising appeals of the evangelical message did much to turn the crushed, bewildered peasants of the Industrial Revolution into the urbane, purposeful potters of today.

The attitude of the parish churches was one of take it or leave it; people were at liberty to find their own way to heaven unless they chose to go to church. Certain it was that the church would never come to them; it was too busy on the grouse-moors or attending to its farms and other such important pastoral pursuits. Wesley preferred to take his religion to the people who otherwise would not have had any, and that is how he came to be preaching in the yard of the Old Foley Pottery. The crowds, in those early days, were so big that he had to preach out of doors, since the buildings were too small to hold them, and when buildings were put up specially for the new sect they were often too small as well, leading to a further form of exclusiveness and discrimination. Within a few years of the heady, golden days of open-air services the followers of Wesley had split up into a small number of squabbling groups, and a new parochialism developed that left many out in the cold who could have done with the comfort and guidance of just simply belonging.

John Wesley had a soft spot for the people of the Towns. He said that, in the Towns, even the poor potters were a more civilised people than the so-called better sort at Congleton.

The period of his ministry saw great changes over the whole of the area. When he first came the industry was still in its infancy, and he thought that salt-glazing, with its poisonous fumes, predominated, though I doubt it did. He found the houses of the towns and villages "mean and poor", and the manners of the inhabitants were no better than their homes. Their pastimes were brutal; he named bull-baiting, cock-fighting and goose-riding. Whatever the claims to the contrary, the potters did not give up these amusements for many years, and they were the subject of some of the earliest of the popular Staffordshire figures, made by artists like Obadiah Sherratt. He was making a bull-baiting group in which the showman exhorts his dog *Now, Captain Lad*, up to the 1830s, and I doubt if he would have made anything so old-fashioned or unrealistic that it would not sell.

Despite the persistence of their primitive pastimes, momentous changes had taken place hereabouts in the period from 1760 to 1830. The houses of the people had changed from two-roomed, thatched bungalows to recognisable town tenements in short rows of terraced cottages. The road system was pretty well established except for the new 'D' road, which has sliced through everything, willy-nilly—and most of the available land in the valleys and on the hill-sides had been dug over or built over.

There is no doubt that the development of the Potteries is due largely to the skill and business acumen of a relatively small number of manufacturers, but they needed some chance, some opportunity, to show what they were made of. Possibly the first of these was the introduction of the oriental habit of hot drinks. This took place gradually during the seventeenth century, and became all the rage during Queen Anne's time, in the first decade of the eighteenth century. Prior to this, local potters had lavished their latent talents on the rather peculiar, charming sprigged and slip-trailed presentation pots, invented in Burslem during the reign of Charles I. Subsequent developments put them in a good position to take advantage of the next fact that emerged, that it was extremely painful to drink hot tea or coffee from a silver or pewter vessel; and experience soon showed that the cups made of wood, horn or leather, which they thought were all right for cold drinks, behaved very badly when filled with scalding hot liquids. They either split, deformed or collapsed making the celebration of common-brew rather like a contest to catch quicksilver. Mulled ale and a peculiar potion called posset were drunk hot, the ale in vast quantities from jugs, and posset through the spout of a teapot-like thing. However, no-one could afford to drink jugs full of tea or coffee; the cost was prohibitive. A whole new range of 'diddy' posset pots with strainers, and tiny bowls for the tea was needed, and the Staffordshire potters set to with a will.

Another Elizabethan innovation, smoking, helped as well. The cigarette had not been introduced in the seventeenth century, and tobacco was far too costly to use in the form of cigars. So this North American weed was incinerated in tiny portable furnaces of clay. Although evidence of the manufacture of clay pipes in what are now the Six Towns is scanty, nearby Newcastle had a flourishing pipe-making industry, and many local lads and lasses must have learned a lot of things that did the potting trade good in the manufacture of tobacco pipes.

So, with all the developments at Burslem, Etruria and Stoke, the Ceramic Industry had grown into a young giant. One of those developments was the bone china of Spode, and it gave the industry a bit of a problem; how to make a profit from this nice new invention, when all their pot-banks were geared up to make anything but. You cannot just change from earthenware chamber-pots to bone china tea pots at the drop of a hat. Today's industry might manage it, but not the industry of the time of George IV, top-heavy with hand labour as it was, and dependent on capricious bottle-ovens.

Next to the Foley was an area that was called, appropriately, Lane End. Like the other parts of the Potteries, it had been extensively mined for a number of years, and by the end of the eighteenth century was beginning to show signs of developing into a town of some consequence. There was room here for the building of more pot-banks wherein to make bone china.

V

LONGTON

Longton is the newest and most isolated of the Six Towns, and is typical of them in that the factories are scattered amongst the houses and shops, and the open country is never far away.

There really is not a lot to say of Longton until 1759, the year in which the main road from Derby to Newcastle, through Uttoxeter, was turnpiked. This road followed beside the Roman road from Tean to Stoke, but was not actually laid on it except between Lane End and Fenton. It may be worth mentioning that a turnpike road was a stretch of road maintained by a small company called a Turnpike Trust. They were empowered by Act of Parliament to raise money to improve the surface and drainage of their bit of road. To recoup their expenses, and make a profit, they charged tolls on all traffic that used the road, controlling the flow by erecting and operating tollgates and they built tollhouses for the people who operated the gates.

What this turnpike trust had done was to improve the lane from the village of Meir to the point where it ended and joined the Roman road. Not unnaturally, the place where Meir Lane ended was called Meir Lane End, later on, simply Lane End. When, in 1771, the road from Stone to Lane End was turnpiked, trade along the road increased enormously and the settlement grew up at the junction of the two roads.

The land has been dug over so much in the last three hundred years or so that archaeology cannot help much to shed light on Longton's past. One large cache of Roman coins has been found recently in the area, though what the Romans did here apart from tramping along their road and burying coins is a matter for speculation.

The earliest mention of Longton was in 1212, when it was called Longeton. The name means a long village, and is a good description of the modern town as it reaches out in three long fingers along its lines of communication.

There is a tradition that there was not much going on in the area before about 1815, but if this is true, one wonders why, in 1760, John Bourne endowed a small house and school at Longton. This was the germ of Longton High School, which went on through various vicissitudes to become one of the great schools of North Staffordshire. The building of a school is always the sign of a community growing to adulthood, especially when, as in Longton, it is followed in a couple of years by the erection of a church.

However, mention of the church is jumping the gun a little. The first building of note we encounter across the border from Fenton is the Boundary Works, on the right. People probably pass it every day without giving it a second, or even first, glance. That is a shame, and their loss, since it is one of the best buildings in the Potteries, and one of the least altered of the early ones, at least as far as the façade is concerned. Built in 1819 it has three storeys separated by string courses. The central part is of three storeys, including, of course, the 'tunnel' and the Venetian-windowed Boardroom, surmounted, not surprisingly, by a pedimented gable. In this case, the gable has half a window in it, as has the rest of the second floor. The second ground floor window to the left of the lovely basket-arch has been turned into a doorway, and the other ground floor windows have had their lower panes bricked-up, probably as a security measure, since the works has been a warehouse for several years. And, I suspect, the making-up of the road to a greater height above 'see level', has effectively brought the windows and the contents of the rooms closer to the prying eyes of the ubiquitous small boys. Despite the minor alterations, on a Sunday morning, when the traffic that roars through Longton on weekdays has slowed to a mild uproar, the ghosts of the place assemble and everywhere is golden and thronged and interesting in front of Boundary Works, if you have the knack of standing, quiet, and looking up. And at other times the ghosts of Longton are missing and it is all barren, however many people there may be in the streets.

The thing to do is to hurry along from here to Times Square, since that is what everyone seems to do. It must have something to do with the railway station, whose presence is announced more by the enormous bridge than by the notices and shabby arches with their alpine steps. It does not seem very likely that the tired old church opposite inspires much urgency.

The original church—or chapel—on the site, was put up largely at the expense of John Bourne of Newcastle who had put up the money for the school on the same site. This was in 1762, and, although it was used for prayer book services, it was registered as a chapel for Protestant Dissenters. This was because it had to have a licence of some sort to be used for services legally, and the way it was done avoided any inter-ference from the riding, shooting, and fishing clergy. It was, however, consecrated in 1764. By 1790 it was too small for the people who wished to use it, a commentary either on their piety or numbers, and was falling down. Not that it had been badly built, it was just that gentlemen would persist in taking away the shiny black rock that was underneath it!

Rebuilding took place from 1792-1795, and the nave and west tower are of that date. The whole affair is of red brick, with stone dressings. The west doorway is typically eighteenth-century Classical in design, but the nave win-dows are slightly pointed, have Gothic glazing bars and are arranged in two tiers. Inside, the original galleries go round three walls of the nave. Supported on slender, elegant cast-iron columns, the galleries are reached by staircases from the entrance porch. The east end was built on in 1827 and makes the church look like a grotesque medieval dwarf with its enormous Gothic hump of steep roofs supported in a bandy-legged fashion by buttresses, the whole accentuated by pin-nacles on the corners like the more extreme affectations of fifteenth-century parade armour. However awkward the additions made the management of the interior (and the roofs do not match), the whole thing has the sort of plug-ugly charm of a deformed but beloved uncle. The Victorian church restorers, who could never leave anything alone, restored this church of St John the Baptist in 1889, but there was little even they could think of to do to it that had not already been done, so they contented themselves with making it safe and strong again, and refitting the interior.

On the other side of Times Square, which used to be called Market Square, and is roofed over in part by the huge, ghastly railway viaduct, is the town hall which, as one would expect, is tentatively Classical at the front which is made of large stone blocks. It has the usual central portico, this time with three bays, and in the upper storey are blind openings with carved panels. In the centre Ionic pilasters support a pedimented entablature. Along the side of the building, in the Strand, is the Market Hall with shop fronts filling the arches of its rusticated walls. Since it has been cleaned the whole building has commanded more attention and is a pleasant surprise after travelling between the tired old walls of the Foley. It was built in 1863.

Past the town hall, Market Street becomes Uttoxeter Road by bending to the right and leaving the course of the Roman road which continues along Sutherland Road for a short way, only to be lost under the buildings. The main road rejoins the Roman one at Railway Road, on the left, but that is of no real consequence, and if we shoot up there to admire that most inauspicious spot, we shall miss seeing what Longton is all about.

Like any other of the Potteries Towns, Longton owes its existence to a combination of geology, geography and a small handful of determined men and women. It is a strange coincidence that the first known pottery in the town that has become the centre of the china industry should have been one that produced china years ahead of its time. This was at Longton Hall, now completely demolished. It stood at the south-east of St Paul's Church and overlooked the Furnace Brook. The works was opened about 1749 and produced a glassy sort of soft paste porcelain. The manager was William Littler, the originator of the brilliant blue ground, known afterwards as Littler's Blue, which the later china factories of

Longton used to such splendid effect on their Victorian table-wares and figures. The body they used was unstable and warped badly in the kilns, and they tended to use far too much glaze, causing the ware to congeal into an unsaleable mass. Everyone associated with the venture lost money, and the firm gave up in 1760. Their experience was an object lesson to North Staffordshire potters who, with the exception of New Hall as we have seen, left porcelain severely alone until Spode came up with something economically viable.

One of the early notables of the Longton potteries was John Turner, whom we have met before. He it was who, in partnership with Banks at Stoke, took in Josiah Spode I, and eventually moved to Lane End to set up on his own, in 1762, somewhere near the Market Street end of Uttoxeter Road, which was then called High Street. He made the usual, money-spinning cream-coloured ware, as well as stone ware, jasper ware and passable imitations of Wedgwood. These last are obviously copies, but often have a much smoother, warmer feel than the Etruria originals.

In common with his contemporaries, Turner used clays imported from the south, but was able to augment his supplies from a deposit of fine clay at Green Dock, an area to the south-west of the town, bordered by the present Heathcote Road. He joined the New Hall Company from 1781 to 1782, and, in 1784, in partnership with Mr Abbott, was styled as "Potters to the Prince of Wales", which is where he acquired the device of the Prince of Wales' Feathers as his trade mark.

John Turner died in 1787, to be succeeded by his sons, William and John. They were as successful as their father, and within fifteen years had two factories on opposite sides of Uttoxeter Road. Old John's time at the New Hall Works had not been wasted. Whilst there, he had learned the secret of Littler's Blue, and the Turners of Lane End became famous for their "shining blue glazed pottery similar to that of the Japanese porcelain".

It may be that their quality and success were their undoing. Pottery had to be good to sell on the continent. Theirs was good and sold well across the English Channel. They neglected their home markets accordingly and, when our blockade of

European ports cut off the continental markets, their business slowly declined and they were bankrupt by 1806. But you cannot keep a good man down, and William was back in business in 1807, back in Longton in 1811, and still going strong in 1830.

One of the best ways to get to know the Potteries is by looking at the product. Any good museum will have an example of Turner's ware, or that of a contemporary of equal quality, and it is most enlightening to see what we were capable of in the days of Napoleon, a long, long time ago.

Another family who helped to make Longton an important pottery centre were the Cyples, who had a lane named after them and a pottery where the Bennett Precinct is now. Both have vanished in the process of redevelopment, though some derelict property of the period still exists [1976] on the fringes.

Longton did not throw up dynasties of potters like the other towns, even though some families had a good run, and many names linger on in modern firms, like the Crown Pottery of John Tams Ltd, who have been there in the Strand since 1875 and Aynsley's of Commerce Street, who have been there since 1873, making earthenware, and who have another factory, the Portland Works, in Sutherland Road, where they make china. The Portland Works is a late version of the classic Pottery style, with archway, Venetian window and pedimented gable, dating from 1861.

Longton, apart from having more actual pot-works than any other pottery town, and more china factories than the rest put together, has also preserved more bottle-ovens. Recent demolitions have opened up the view of the old potteries very well, and quick diversions up side streets for a peer-about can be most rewarding. Whole streets of terraced houses, shops and small potteries have vanished, making it rather difficult to see the surroundings of town life even a few years ago. The people have moved to smart new housing-estates nearby at places like Blurton, Dresden and Weston Coyney. The shops have moved into the precinct and the pot-works have been rationalised.

Vistas that have not changed much are Uttoxeter Road, Normacot Road and Sutherland Road. Walking in these roads we get a good impression of what the town was like at the end of Victoria's reign.

Uttoxeter Road has become the scene of the most remarkable and spectacular feat of conservation in the country. This is the Gladstone Works, chosen as the home of the Living Industrial Museum of the Potteries.

I have already advocated looking at pottery and walking around the Towns as two of the best ways of getting to know what makes the place tick. The one, very best way of making that worthwhile discovery is to visit the Gladstone Pottery Museum, winner of the 1976 Museum of the Year award.

The works is on the right of Uttoxeter Road, just past the Union Hotel, on the corner of Chadwick Street. Its history covers the whole span of the industry as we know it today and reflects the development of Longton in a unique way.

The biggest land-holding in the area was the Longton Manor Estate, which had been in the hands of the Foley family from Birmingham since 1651, one of them became Baron Foley of Kidderminster in 1712. This family's name is that of the area between Lane End and Fenton. The Foleys held on to the estate until 1773, when they sold it to Obadiah Lane, a parson, who was the sitting tenant. Strangely enough, the Lanes, who were Potteries folk, followed the Foleys to Birmingham and the house was subsequently sold, in 1775, to a Thomas Fletcher. The rest of the estate was split up into small parcels during the 1780s, enabling enterprising potters from the other built-up towns to set up compact, efficient businesses which were soon turning out the new miracle of the age, china.

One family of well-known local potters who bought some of the Longton land were the Shelleys, who got a good plot with a frontage on the new turnpike road at the southern end of Lane End. The Gladstone Works is located on part of the site. The Shelleys bit off more than they could chew. They not only made their own high-quality earthenware, they bought Wedgwood 'in the white' and decorated it themselves, which must be a hard way of making a profit. By 1789 they had become bankrupt and sold the land, with its buildings, to William Ward for the then large sum of £900.

He had more horse-sense than the Shelleys and split the site up into two more manageable portions, each making, even so, a good sized pot-works. One part became the Gladstone and the other the Park Place Works. Visitors will have no trouble seeing how the pieces fitted together, and one cannot fail to be impressed with the ambition and size of the Shelley holding.

The subdivision did not end there. The bank was split up into a number of even smaller lots, each tenant putting up buildings of one kind or another. Some of those old walls still stand, incorporated into the fabric of later buildings, and everywhere you are surrounded by little pieces of the eighteenth century.

When, in 1818, the works was sold to John Hendley Sheridan, it was typical of the smaller concerns of the day. The usual scheme, which was followed at the Gladstone, was for the master's house to be built conveniently close to the road, with a kiln out the back and a range of workshops around the edges of the property. These little shops were rented by working potters who either made ware of a specialised nature in a small way, or decorated the products of other potters 'by the piece'. These out-works were a common feature throughout the long history of the industry. In fact, there are still firms with names like the Perfection Porcelain Company, who, if you visit them, turn out to be two very old fellows named Ern and Lishe, and a very pretty girl whose name is Nellie. They work in a small, dark shed, decorating white china seconds with exquisite care and firing it in a large muffle kiln that looks as though it might have been the bake oven from the Ark.

Meanwhile, back at the Gladstone, Mr Sheridan had found himself a tenant who liked the way the back-yard workshops had grown, and also the fact that extra kilns had been erected. This was Thomas Cooper under whom the works took on its present form. He was in the business of making china table-ware and Parian figures. Parian ware was developed by Copelands and, when well made, closely resembled the marble after which it was named. Soon, potters everywhere were knocking-out superb copies of classical and neo-classical sculpture in the form of statuettes. Not the least of these was

Tom Cooper, who, by 1853 had bought the Master's house in Chadwick Street, and followed this by demolishing the two old houses in the High Street (Uttoxeter Road), replacing them with the buildings more or less as we see them now.

Liverpudlian merchants are always popping up in the Potteries, and this time it was the son of one, William Ewart Gladstone, who, in 1863 came to lay the foundation stone of the Wedgwood Memorial Institute. His first important speech was in defence of his father whose merchandise was often Negro slaves, and who ran a slave plantation himself in Demarara.

It is rather strange that the kind people of the Potteries should have one of their works named after a supporter of slavery, when a popular smoke of the period was a terracotta tobacco pipe with the tiny figure of a curly-headed negro kneeling before a frock-coated man whose foot is on the negro's neck, and printed along the stem are the words "It is shame to treat a brother like a dog". But the factory was named after him, and, I suppose "a rose by any other name would smell as sweet". A long succession of owners made china on the site, right up to 1960, when Gladstone fired its last bottle oven. It may well have been the last such firing in the Potteries since by that time the old kilns were being phased-out, but there are other claimants for that particular honour, and I am not keen to be the umpire.

The run-down was gradual; the firm—Thomas Poole and Gladstone China—continued to decorate their previously-made ware, and when that was all done, they used the premises as a despatch depot for what remained of their stock. Then the works was put up for sale, and no one wanted it.

The end of that part of the story is rather like Haydn's 'Farewell Symphony', where each player blows out his candle in turn, and quietly leaves the stage.

Like many another venerable pottery, Gladstone was threatened with demolition. That was where the 'Uncles' stepped in again. They, together with some other uncles from museums, had been going quietly round the Towns to see what there was that could be saved that was worth saving. There was much they would have liked to have saved, but pennies were in short supply, and the bulldozers moved in and gobbled up one after another.

In March 1971, the bulldozers were closing-in on Gladstone, when an Arch-Uncle named H. and R. Johnson intrepidly halted the gnashing car-park makers with the money to buy the site, complete. This act of incredible public-spirited benevolence was a real case of "physician, heal thyself". The local authority could not provide this great national industry with the wherewithal for an Industrial Museum, so the industry did it itself, in the form of one of the world's leading manufacturers of glazed tiles.

At that stage, the idea was really to save the building as an example of a typical Victorian Pot-bank. A Trust was formed to plan and administer the venture and a group of eminent pottery manufacturers had come to realise that the pottery could house a working museum; a considerable advance on Ironbridge and Cheddleton, admirable though they are.

So a Living Museum was born: one where visitors could see the pottery manufacturing methods in an authentic nineteenth-century pot-bank. It speaks volumes for the enthusiasm of the industry and their friends that the Staffordshire Pottery Industry Preservation Trust had managed to raise over £160,000 within twelve months of starting.

Not that the job was easy. The small force of professional and volunteer workers who set to to put the aims of the Trust into practical bricks and mortar, soon found that the buildings were in a far worse state than had been suspected at first. This meant that large sums had to be spent on the fabric; money that had been earmarked for the construction and furnishing of galleries for historical displays.

It is little short of a minor modern miracle that the restoration and conservation of the Gladstone has been so authentically and accurately carried out, when the original condition of the site is considered. The number of variables in the equation made the job very tricky. Not the least of the problems was the need to make it perfectly safe for visitors, since they are the ones who matter, without spoiling that atmosphere of rickety stairs and dank, dark corridors that pervades the older section of the pottery industry.

I do not know what subtle witchcraft they employ, but the firm, safe stairs and clean, bright corridors still manage to give the feeling that remains in some of the derelict, unrestored works.

The plan of the museum is very simple. The ground floor is a working pottery where pots are made as they have been made for over one hundred years. Upstairs there are galleries telling the story of the development of the industry from its most distant origins up to the present. A principal theme is the rise of the Staffordshire Potteries.

The greatest attribute of this fascinating place is hospitality. Visitors are made to feel welcome right from the start, and the welcome never wears out. This is not one of those museums where everything is locked away in glass boxes. In most cases you can walk right round the exhibits and quite a few are touchable.

Although the potters working there are quite busy making rare items for the museum shop, they always have time to explain what they are doing, and if visitors are really keen, they are sometimes allowed to have a go.

For the very enthusiastic there are classes in various aspects of potting, and there are archives that are rapidly reaching legendary proportions. On top of all this, the Gladstone is host to a vast range of activities, social as well as historical. On at least two long-drawn-out occasions when any other museum would have ejected me as an infernal nuisance, the over-worked staff have patiently thrown open their resources. When making a series of radio programmes, I trailed a group of lively ten-year-olds interminably round the works but did not seem to get on anyone's nerves. Actually, that episode exposed what, to me, is the only flaw in the authenticity of the Gladstone. The children enjoyed themselves thoroughly, which is a good thing, but they all thought that the pottery must have been a nice place to have worked in. Perhaps the creation of the interminable, soul-destroying drudgery of a Victorian pot-bank is not the responsibility of a living museum, but I hope that, one day, a chamber of horrors can be installed, so that we can learn to value ceramics in human, rather than industrial terms.

A visit to the Gladstone is always welcomed by my full-time students, and catering for such parties is all part of the normal, everyday functioning of the museum, but an example of their devotion "above and beyond the call of duty" was shown when I was seeking a venue for an industrial archaeology working party who could only meet when normal people were putting their feet up for a well-earned rest. True to their ideals, the museum staff opened the door and let us in, gave us access to their priceless archives, machinery and industrial hardware, *and* made us a cup of tea.

Most museums are good for only one visit. Not so the Pottery Museum. Apart from the ever-changing exhibitions, like the one of the costumes and so forth for the television version of Arnold Bennett's *Clayhanger*, for which Gladstone provided a good deal of information to make it authentic, there are also permanent displays of the industry. One of the most powerful and impressive exhibits is the Tile Gallery, which takes us right through the history of this useful and often beautiful product. Most visitors leave that room regretting that modern décor has relegated such lovely things to lavatories and subways. Our ancestors had more sense.

Lavatories are well represented in the fascinating Loo Gallery where the gaily painted flowers, written ornament and names like the Thunderer make us wonder whether the Victorians were as dour as we have been led to believe. It also serves to remind us that the ceramic industry provides articles of convenience as well as beauty and utility.

The Gladstone Works is appropriate, since it does work, very well. Here it is that the uninitiated can understand the workings of those machines that gave the potters of the past such a bad time.

As is only right and proper, the whole place is given a beat and pace by the steam engine that would have provided the power for the whole place in the past. The steady clack-pause-for-breath-clack of the engine is one of the first impressions on entering the tunnel, and the orange of the fire from the engine house across the yard looks warm and inviting on a November visit in the fog. But you will not get very hot by that glowing furnace, and anxious mums need have no fear, young Sam can touch the fire as much as he likes, the flames are make-believe. They do not fuel the steam generator to power the reciprocating monster by the wall. It is run on compressed air, a stroke of genius in an authentic setting. There is a delightful feeling of trespass in there, as though you have caught the engineman out whilst he has gone to retrieve his newly-heated pie from the bottle-oven in the yard.

A real bottle-oven. You can even go in, and discover that the big brick bottle is merely a massive masonry mantle for a cylindrical oven inside. That is where the pots were stacked in their saggars. The proper name for the brick bottles is kiln, and this type consumed mountains of coal, and produced dense clouds of smoke. They normally took from eighteen to twenty-four hours to fire, that is to complete the cycle of operations that included filling up the doorways with firebrick and clay, lighting the fires, heating the ware slowly up to the temperature at which they would be done, and cooled down to a temperature at which men could go in and 'draw' the ware.

On a well-managed pot-bank, one kiln would be firing while the other was being filled by the placers. Since productivity has always been important, the process was hurried as much as possible, so that the kilns would produce the maximum amount of ware in a given time. The achieving of this speed-up led to one of the worst evils on the pot-bank, that of drawing the ovens when they were hot. They were drawn when they were so hot that it was not safe for men to be in them. Anyone who refused to go in was sacked and there was always another ready to take his place. They covered their heads and bodies with thick cloth, flannel if they could get it, and had huge pads of cloth on their hands, for the saggars had to be carried out. Their breath was stifled by the heat and the sudden vaporising of the moisture in the cloths. Men were said to suffer "extreme torment"; their ears began to smart, the pain under the finger nails was excruciating. This was not all: they felt as though their noses were bleeding, as though they were being blinded. The men lost so much bodily moisture through perspiration that they dehydrated and became very ill. They did not live to make old bones.

Machines that can be seen inside include the throwing wheel, on which the clay is formed under the hands whilst rotating on a horizontal table. This device, the one that made pottery of high quality and uniformity available, is now obsolete in the modern industry, having been superseded by casting in some cases and by press-moulding in others.

Two other gadgets that ousted hand-making methods are the jigger and the jolley. Both of these can be examined at Gladstone; they were the ones that caused so much trouble in the past.

A jigger originally meant any mechanical aid to producing a constant stream of identical articles, and the word is still used in its reduced form, jig, by the engineering industry and usually means a kind of clamp to hold something being worked on.

To produce plates and other 'flat wares', the potter uses a spinning 'table' with a solid plaster mould, made in the shape of the inside of the plate, fixed to it. On to this mould he puts a 'bat' or slab of clay, then, to shape the underside of the plate he presses a profile template down on to the clay whilst it rotates a few times. The mould, complete with its new plate, is then removed to be dried, when it will come away from the plate which is then fired, decorated and glazed.

The potter calls this process jiggering, a word that comes from the clamp which serves both to hold his former, and to prevent it from descending too far on to the bat and making it thin.

The other mechanical aid to the making of pots is the jolley. When first introduced, any machine for forming pots was called a jolley, and presumably acquired the name in the Northern potteries of Leeds or Newcastle, where machinery was in general use before it came to North Staffordshire, which at that time was rather backward in some respects.

The jolley is a device for forming the inside of deeper vessels like cups or bowls, and, like the jigger, has a small table that can be rotated, usually horizontally. A plaster mould in the form of the outside of the cup is put on to the turn-table, a blob of clay of the correct size is put into it and a former having the profile of the inside of the cup is pressed down into it, the rotation forcing the clay up the mould to form the sides of the vessel. Machines like this can turn out many thousands of perfect pots every day, and were a serious threat to the throwers who could rarely, if ever, keep up with them.

Descriptions of three dimensional processes are never very satisfactory, so it is a good thing that the Gladstone Museum exists to demonstrate and explain them with great patience and tact.

The machines at the production end are only the tip of the iceberg as far as mechanisation is concerned. They were also the ones that annoyed the operatives most. There does not seem to have been any serious opposition to steam engines which must have helped their earning power, nor to some of the machinery that took over the very hard work.

The heaviest labour in the pottery industry was always associated with the preparation of clay for the potter, and the placing of the ware into the kilns after it had been made. In the latter case, a few ounces of unfired china figurines might well be contained in a saggar weighing a stone or more.

At the other end of the production line, all the crushed-up particles that went into the recipe for 'clay' were mixed with water. In the early days the job was done by hand, men and women laboriously stirring monster puddings for hours on end, until the consistency was just right. This mixing was called blunging, a super word that lives on locally, not only in the pottery but also in the kitchen, where a mixed grill, or something very like it, is known as a blunge-up. The creamy mixture that the blunger produced could either be used as it was for casting in plaster moulds which absorbed the excess moisture, leaving the mould intimately lined with clay; or it could be thickened up by evaporation of the water until it was the right consistency for the potter. The thickened-up clay usually contained a good deal of air that was about the same as gunpowder in the kiln, and this explosive menace was removed by 'wedging', in which young, strong, healthy people lifted large lumps of 'body' high above their heads, to bring it down with a hard thump on a solid table where they leaned

on it and pressed it, both processes serving to remove the air:
The lumps of clay would be periodically inspected by being
cut up with a wire, rather as a piece of cheese is, until all trace
of air was gone.

That process, too, was eventually replaced by a machine,
called a pug-mill, in which either a plunger or screw thread
squeezed the air out of the clay under many tons of pressure,
and shoved it out through a nozzle in a convenient shape to
be cut into bats for the potter.

One of the most unpleasant jobs was in the boiling rooms,
where the clay was thickened up. This was replaced by the
invention of filter presses in the 1850s in which the liquid clay
is fed into hollow trays lined with cloth. The slip is pumped
into the trays under great pressure, the water straining
through the cloth, the clay remaining behind. The process was
simple and effective, but because the filter cloths were of such
high quality, as was the wood-work of the trays, it was much
more expensive than the boiling vats, where the only cost was
the labour involved and coal for the fires which was very
cheap. Nonetheless the filter-press, as it came to be known,
gradually took over from the open cauldrons, partly because
pottery manufacturers cannot resist new toys to play with, and
partly because clay produced by filter presses contains hardly
any air and needs very little wedging. Despite all the clever
inventing and hard work that has gone into the machinery,
there are potters today who claim that pug-mill clay is not as
good as wedged clay, and most people would prefer hand-
thrown pots if they could get them, and if, in addition to their
originality and spontaneity, could be added fineness of the
industrial article.

Not that all this machinery superseded hand skills over-
night. There were plenty of things that could not be made as
well by machine as by hand. The old firm of Henry Tams
who used to have a works near the old post office in Chancery
Lane, made coloured marbles, canary eggs, pigeon's eggs, duck
and hen's eggs, birds' nests and linings, and Spotted Jacks,
whatever they may be. Local legend has it that the hen's eggs
were realistic enough to fool not only broody hens, but also
ravenous husbands, and others. The factory was closed at the

end of the Second World War, presumably when battery boxes replaced broody hens.

Other products have damaged the ceramics industry, of course, notably plastics during the period when it was being used to copy everything. But now that it has assumed its own, useful identity and is being allowed to do its own thing, a streamlined pottery industry has found that peaceful co-existence is not too bad.

There have been many hard times for the Potteries, sometimes caused by competition from foreign industry, but usually acting as a barometer for the nation's economic climate. When things were bad, they were very bad. Even in easy times the labour-rich industry was hard to get on with. In the general recession in the period between the wars, many tiny works closed for ever, and some of the larger ones worked a sort of unofficial short time.

Workers would crowd into the tunnel in some works, the yard in others, hoping to be 'set-on'. Amongst the paintresses, a young, pretty girl could usually get a place, according to some who neither young nor pretty, and occasionally a mother whose children were hungry would come with a basket saying, "Could your missis do with a dozen eggs master?"

Even the churches or chapels the workers attended could influence their employment prospects, as many religious establishments were supported by wealthy manufacturers who liked to see the pews filled on a Sunday.

One of the real benefits of the demolition programme is the present appearance of the Church of St James the Less, a most attractive example of a building in the lovely golden Hollington stone, chosen many years later for Coventry Cathedral.

Built in 1834, the church is a good example of Victorian Perpendicular. Apart from the colour, the most striking features are its west tower ninety feet high, and the clerestory windows of the nave.

The church lies between Uttoxeter Road and Normacot Road. Directly opposite is Webberley Lane, where there is a fine jumble of Victorian school buildings now occupied by the

Workers' Educational Association, which is a most influential force in the lives of many local people.

The buildings are known as Tawney House, a tribute to one of the great local patrons of education. It was the North Staffordshire W.E.A. who, as hosts to Francis Klingender at Barlaston Hall at week-end and summer schools, provided much of the material and discussion that led to the writing of Klingender's classic *Art and the Industrial Revolution*.

There cannot be any area in the world in which the Industrial Revolution and art are more inextricably mixed up than in the Staffordshire Potteries, yet it failed to produce the large canvasses that have immortalised other locations whose period of industrial activity was far shorter.

We must content ourselves with the exquisite miniatures that appear on some of the nineteenth-century pots. Perhaps like William Blake, a contemporary of theirs, the painters and paintresses of the early days of the last century could see Heaven in the wild flowers they painted with such care and accuracy. When they were copying Chinese and European designs their work was always stiff, often ridiculous. When they were painting the flowers of the hedgerows amongst which they lived, their work is fluid, sometimes sublime.

One of the places in which they learned to manage the difficult techniques of painting on china, and to distil those syntheses that brought a harebell to life in cobalt oxide on a clay plate, was the Sutherland Institute. Opened in 1899 in Lightwood Road (then called Stone Road), the Institute is still in use for educational purposes, and houses a large public library. It is opposite Webberley Lane which hems one side of the Recreation Ground made as an unemployment relief work in the 1920s. The 'Rec' is on the site of an enormous pit-heap that was levelled by out-of-work miners and potters using only hand tools and wheelbarrows. The shafts of the mine that produced the heap are now buried under a garage and private garden on the east side of Lloyd Street.

We have now moved out of Longton into its suburb of Florence, named after the eldest daughter of the third Duke of Sutherland, one of the nation's largest land-owners, and certainly the owner of the most land in that area. He and his

father were responsible for most of the streets in greater
Longton, and put up a large number of the houses that remain
from the 1860s and 1870s. The Duke of Sutherland also gave
the lovely park to the town. Known as the Queen's Park in
honour of Queen Victoria's Jubilee, it was opened to the
public in 1888.

Compared with other land-owners, the Dukes of
Sutherland treated the people on their land quite well. They
threw open the grounds of their home, Trentham Hall, on a
day in the summer known as Trentham Thursday, and, in
addition to donating the park, gave the land for a number of
schools and other buildings, as well as chipping in some of the
cash for the bricks and mortar. This is all the more surprising
when one appreciates the extent to which the good folk of
Longton contributed to the discomfort of the Duke and his
household.

The contents of the privies, cess-pits and drains of Longton
were emptied into their brooks and streams, as was the custom
of Towns of the day, and the streams emptied into the
Longton Brook which disgorged into the Trent right alongside Trentham Hall.

The Longton effluent was not the only noxious substance in
the Trent, but it was the last straw. The Duke did his best to
alleviate the problem by seeing to it that proper sewers were
built in the terraces he was developing, and gave up an
enormous 250 acre site at Blurton Waste for a sewage farm.
This must have helped, but it was not enough. As the new
twentieth century crawled along to its first decade the prospects of a Federation of the Towns seemed rather dim. Only
concerted, unified effort could solve the problem of sewage.
There was only one way it could go if proper disposal works
were not built, and that was into the Trent and past His
Grace's house where he was wont to entertain royalty and
other people with delicate noses.

By the time Federation seemed to be a possibility, it had all
become too much. The Duke decided to leave Trentham and
offered the house to the newly emergent County Borough as
a centre for advanced education. They knew about the smell
and flies however, and turned the offer down. The house was

demolished, except for a portion of the stables, and the gardens were turned into a pleasure ground. If old photographs are anything to go on, the Towns missed a beautiful building, and it was not long before joint action improved the Trent enough to make Trentham Gardens bearable, although the river is still one of the worst in the country. The gardens are internationally famous and have everything to offer for a day or evening out. They are still overlooked by a massive statue of the Duke, which is supposed to mark the spot from which everything he could see was his.

He could certainly see his lands in Longton, amongst which was the ancient township of Normacot, whose name means 'Norseman's Cottage'. Appropriately, Normacot had a cottage hospital. This had a number of homes before the present one, in Upper Belgrave Road, was built in 1890 on land given by the Duke of Sutherland.

There was no National Health Service in those days, and hospitals were financed by a mixture of benefactions and public fund raising efforts. One such fund-raising effort was associated with John Aynsley who started life in a humble way on a pot-bank, and went on to be the owner of an important china factory, and became Mayor of Longton. He was one of those rare spirits whose vitality and wit transcended the dull and dirty days of the Potteries to make life sparkle for a while for anyone who came into contact with him. He wagered a large sum that he could drive a pig from Trentham Gardens to Longton. The losers were to give their money to the hospital. The Duke was one of those most heavily engaged in the bet. As cheering crowds thronged the roads, John Aynsley duly drove the pig to Longton Town Hall and the funds of the hospital were suitably augmented.

As the result of another wager Aynsley came into possession of a donkey. He called it a Jack Ass. This animal too was destined to help his pet project, the Cottage Hospital. He conceived the idea of raffling the unfortunate beast; the proceeds to go towards the hospital fund; the winner to hand it back to be raffled again. This went on for a long time, I hardly dare say "donkey's years", and eventually the donkey died, as even donkeys must. That was not the end of its

Real bottle ovens, the calcining kilns of James Kent

Fenton Town Hall

The most famous chimneys in the world – the bottle kilns of the
Gladstone Potteries

Longton preserves more of the industrial past than the rest of the six
towns put together

Longton Town Hall

PETER DOBING

The Longton skyline – church and chimneys

Newcastle's Guild Hall, with stalls of the ancient market at its feet

St. Giles' Church, with its neighbour, the Old Meeting House,
supervises a busy traffic island on the A34

The Old Orme Boys' School, on the site of the town 'lock-up'

Tunstall Town Hall

Harecastle Tunnels: Brindley's on the left, Telford's on the right

services to the hospital. The long-suffering animal's skin was tanned and sold, the meat went to a cat's-meat shop, and the bones were calcined and used in a china-body to make a model of the donkey which stood in the entrance hall of the hospital, over a collecting box. This box was in use right up to the time when the National Health Service took over the financing of local hospitals, and in the receipts column of the ledger was always a line saying:

"Donkey....£ s d"

The area of Longton in which Queen's Park is to be found is Dresden. Careful enquiry has failed to reveal any Duke with an eldest child called Dresden, so it is supposed that the suburb was named out of respect for one great china town by another.

Much has been made of sewers already, but the preoccupation of the residents is understood when plagues of only one hundred years ago are considered. We take our drains very much for granted today, but the Victorians, who had a great sense of occasion, treated them with great respect. During sewerage work in Dresden in 1976, workmen opened up a manhole in one of the original, brick built sewers which was being replaced since it was very near collapse and had been severely affected by subsidence. Inside, they found a memorial tablet, measuring two by two and a half feet. At the top it said:

> This memorial stone of the Sewerage Works in the Borough of Longton, was laid on the 31st March, 1875, by Aaron Edwards, Esq. Mayor.

It goes on to record the names of the Town Clerk, the borough engineer, the surveyor and the contractors. Discovery of bricked-up sections of the old sewers gives rise to excited little stories of secret passages and buried treasure from time to time, and their devotees are as welcome to their fantasies as are those who search for the crock of gold at the end of the rainbow.

In fact, these old sewers do sometimes give up interesting information. They were provided with settling pools and carrion traps. What was ever caught in the carrion traps I would not care to guess, but lots of little goodies ended up in the settling tanks, articles lost down the wide throats of old brown slop-sinks that always smell of soft-soap and carbolic.

Small armies of sewer men worked in those old drains and culverts, often subject to the same dangers as miners. There was always the chance of a flash flood caused by rain miles away when the skies were clear overhead. The slowly-moving sewage of a town poorly supplied with running water generated large volumes of gas which was both poisonous and highly explosive. To detect the gas the sewer men carried cages of canaries, whose tiny lungs responded to the deadly fumes long before the great organ-bellows in a sewer man's chest even began to wheeze. Subsidence, unsuspected on the surface, could make holes hundreds of feet deep in the sewer floor which was nasty for 'Dan' as he was nicknamed, however many of him there were, and the holes were expensive since the sewer had to be opened up from the surface and tons of shraff were tipped in.

This tipping of shraff was an important business. Considering the vast output of the potteries, even a tiny percentage of waste would have produced mountains of broken pots and added to this were the used plaster moulds. The moulds did not last long; they soon became impregnated with clay, which spoiled the crispness of the finished article, and, being fragile, they were often broken and chipped. Fortunately, the digging of clay provided big holes for the waste to be dumped in. Such dumping was a skilled job. It was also dangerous, as most jobs seem to be in North Staffordshire. The method usually adopted was this: a high-sided cart was piled up with the shraff and driven off behind sweating horses towards the nearest tip. Such places were sometimes on the premises of very large firms who dug their own saggar marl, but normally the smaller firms took their rubbish to special tips rented out by the owners of the land. The horse was made to back the cart towards the edge of the pit and the driver, standing on the load, started to shovel the load off down the slope. Each of the towns has a story of the horse, cart and driver who backed too far down the slope and vanished into the muddy water at the bottom. Future archaeologists should have an enlightening time when they delve into these treasure pits, especially since much of the ware that was dumped would not look too bad to the average observer. The standard of inspection in even quite modest potteries is very high, the skilled eye and trained fingers detecting minute flaws that would eventually make an article unsuitable for the purpose for which it was intended, even though it might look all right to ordinary mortals. In this department as in every other, the Potteries had its legendary heroes and heroines. There was a freelance china 'looker-over' named Annie Fringe, who lived the life of a tramp in the Towns, but who could always get work in any china factory, so great was her skill. She was last heard of leaving for South Africa, where she was going to grade ostrich shell, used by some west-coast tribes for jewellery.

A local hero with a distinctly non-ceramic background was Reginald J. Mitchell, born at Butt Lane, which is part of Newcastle-under-Lyme; he grew up in Longton. His father was a school master, and they lived in the school house, nearly opposite the Church of the Holy Evangelists, Normacot. The house remains, but the school has gone. Perhaps the environment influenced his choice of career, as, behind the churchyard, by the vicarage, stands the lovely old building which was once the Furnace Mill. The stream that powered the mill still runs out of an iron spout nearby. That same stream served the furnace that gave the site its name, one of the oldest such furnaces in the country, and, although the place has now been levelled and is used as a playing-field, there must have been plenty of evidence of the old works there when Mitchell was a boy.

Whatever it was that influenced him, young Reginald turned to engineering, which was nearly as important locally as potting. He was an ardent maker of model aircraft as a boy, and was educated at Hanley High School. After school days, he was apprenticed to Kerr, Stuart and Company, engineers, of Fenton. He was a student of the old Technical College, in College Road, Shelton, which became the Polytechnic. There was no local firm to satisfy his love of aircraft, so, having served his time, he went to work for Supermarine Aviation, in Southampton. They were largely engaged in making huge seaplanes for the Royal Air Force.

By 1919, at the age of twenty-four, he was their chief engineer, and was involved in the prestige project of building seaplanes for the Schneider Trophy air races. After their third successive win in 1931, with the then incredible speed of 340.08 miles an hour, Supermarine won the trophy outright for Britain. Experience gained in producing aircraft for the races encouraged Mitchell to enter the competition to produce a new fighter for the R.A.F. The aircraft that resulted from his efforts was the Spitfire which was first flown in 1936, and which played such a decisive part in the war in the air in the 1940s. Beautiful to look at, and perfect to fly, the 'Spit' was typically a product of the Potteries' brain, combining as it did perfection of form and excellence of engineering.

All of which helps to explain why there is a flourishing Model Aircraft Society called the Spitfires who fly their aircraft locally, and why there is a Mark 16e L.F. Spitfire of Normandy-Landing vintage in a plastic bubble in Hanley.

Also from Longton was one of our most enigmatic Englishmen, Havergal Brian. Born in Ricardo Street, Dresden, in 1876, he lived for his first few years within spitting distance of where Queen's Park was going to be, since, although he only died in 1972, at the age of ninety-six, he was already eleven years old when it was opened. Growing up in such a very confident, Victorian town, it is hardly surprising that he developed his immense talent into a series of Gothic master-pieces of Wagnerian proportions. Ignored by his contem-poraries except for a loyal, perceptive few, he must now rank with Mahler and Holst. It is strange to imagine this friend of the Arch-Victorian, Elgar, surviving into the era of supersonic transport and men-on-the-moon. Old age was not common in his generation; poverty, neglect and industrial disease cut short the lives of too many of his fellow citizens. Musical and optimistic to the last, how they would have applauded both his symphonies and his posthumous success.

VI

NEWCASTLE-UNDER-LYME

When Dr Plot visited North Staffordshire, he referred to both Burslem and Hanley as being near Newcastle-under-Lyme. Nearly a century later, Doctor Pocock wrote and told his mother that Newcastle on Lyne, as he called it, was the capital of the Potteries villages, and that it had a handsome church and a market house.

The name is the first thing that interests us in Newcastle. Colloquially it is known as 'Cassal'. The earliest documentary reference to the place is in some papers known, rather splendidly, as Pipe Rolls, dating from 1168. At that time it was called *Novum oppidum sub lima*, which means 'The new castle under the elms', but the word *oppidum* used for 'castle' is the name the Romans had for an ancient British earthwork and there is no trace of any such fortress in the immediate area.

A few years later they were using the more normal word *Castellum*, which is easier to translate. In 1317, the good folk of the city of London, in their letter-books, had changed it. Now it was Newcastle Super Are. Meaning Newcastle above the River Are. This has echoes of 'Normacot', because *ar* is the old Viking word for River, and we know from their poetry that it was pronounced *ayr*. There is other evidence of the Norsemen hereabouts, and one of them named Knut, which we in our history books write Canute, had a place to live just outside Newcastle; it is still known as Knutton.

Really, the modern name has the Lyme in it that we saw at Burslem; it is a new castle, built outside the great Elm Wood. The Lyme Brook was named after the town, not the other way round. I wish they had stuck to Newcastle-on-Ayr; it has a sort of promise and charm that is much more attractive than the charnel-house undertones of being under lime.

There is not much point in looking for the castle. Most of it was gone by the 1540s; all that then remained was "one great towre" and the rapidly-silting artificial lake that had been made as a protection for the original Norman keep. The lake had been formed by damming the brook that ran through the valley. The dams are still remembered in the names Pool Dam and Rotterdam. The pool itself had degenerated into a marsh in the centuries between its creation in the early 1100s and 1828 when the pool was drained, leaving only the Lyme Brook as a memento. The extent of the water can be seen in the school playing-fields which cover most of the area of the old castle pool. The only other tangible remains of the building that gave the town its name are the remnants of the motte, or mound of earth thrown up by the Normans, at the north-western end of the Queen Elizabeth Gardens, and the remains of the foundations of the great gate house in John O'Gaunt's Road. Apart from this we have names like Upper and Lower Green, Holborn and Bridge Street, to remind us of the fortifications that were rendered useless by the vulnerability of its low-lying position to the cannon introduced into warfare at the end of the Middle Ages. Quite a lot of substantial stone work has been uncovered in the area during excavations, and the museum has some timbers and other things recovered from the site, as well as good maps that help to explain where everything used to be.

The first impression of Newcastle is that it is a slow-moving, pleasant town, with a number of interesting-looking buildings. Unfortunately, the needs of modern commerce, road building and the public health acts have, during the post-war period, removed much that was worth looking at. Even so, there remains plenty for the sightseer and collector of pleasant buildings.

Of the garrison town that clustered round the gate of the castle, nothing remains but the names of its streets. The English, under their one and only dose of foreign domination, seemed to develop the peculiar knack of living quietly in the most appalling conditions of damp and squalor. Once they had come to terms with the Normans, however, and turned the episode to their own advantage by claiming to own France by right of having been conquered, they soon moved to more comfortable quarters.

The castle had been built there in the first place to keep down a particularly mutinous group of Saxons who had to have their homes burned before they would accept the Verdict of Hastings, and the site chosen controlled the 'T' junction where the road to the west branched off from the road to the north from London. The North Road ran along the foot of the ridge that bordered the castle pool and formed the valley of the Lyme Brook.

Once the Saxons had absorbed the Normans and recovered their dignity by becoming the English, they moved their homes away from the main road and the deafening plod, plod, plod of the ox-carts. They went up on to the ridge, above the church that supervised the activities of the lake in the valley. There they laid out their town around the rectangular space needed for the markets and public near-orgies of which they are still so fond.

The pattern of the streets of the town is still dictated by that old layout, the main thoroughfares being connected by a number of alleys at right angles. When the pattern becomes clear, moving about the town on foot is surprisingly quick.

It is a well-known fact that in those early, feudal days, people held their land by paying with their time and service; sometimes with their lives. What is not often appreciated is that it was possible to have an altogether different arrangement. In the feudal service deal, the serfs were under the thumb of either a bad, bold baron, or a mighty, merciless monk, but both the baron and monk were, to some extent, in the power of the monarch. The barons were supposed to help the King with money, soldiers, home-grown produce and moral, or immoral, support. Quite often the money was slow in filtering through into the King's coffers, if indeed it ever arrived. The soldiers took many wrong turnings, there were notoriously bad harvests and the barons and monks found other causes to shout about. All of which left the King short of everything he thought he needed, and probably did. One trick the monarch had left was to take the land from its original owners and let it out in small parcels called burgages to people who were, thereafter, called burgesses, and the settlement created was known as a borough. In Newcastle, the regular, rectangular plots divided up by roads, lanes and 'entries' are the physical remains of the burgages.

The original burgesses of Newcastle-under-Lyme paid a shilling a year each for their burgages, had a share in farming the open fields, and, best of all, were let off every form of service to the manor. They could get on with the business of being commercially successful by charging duty on all goods that used their crossroads, and by holding regular markets in which they not only sold their own goods, but also levied taxes on all who came from outside to sell theirs. Newcastle has prospered quietly all the way through its history, and has enjoyed for centuries those precious rights of self-government as a borough that their wealth enabled them to buy from the crown.

The history of Newcastle, then, is very different from that of her neighbours, and the "Loyal and Ancient Borough" has always stood aloof from her grimy, hard-working sisters. When the Potteries were mere clusters of crude hovels, Newcastle had fine houses and was represented in the affairs of the nation by Members of Parliament. As the Six Towns grew, they made a number of hard tries to 'get' Newcastle as a satellite. The citizens of 'Castle have successfully resisted every attempt to take them over, their most convincing argument being that they are 'different'. The borough now embraces the city in a wide, crescent-grasp that reaches from north to south and covers an area roughly equal to that of all the Towns put together.

This vast area is the result of the reorganisation of local government in 1974, but reflects, in part, the old manor of Newcastle, ruled from the castle, and separate from the town. It originally included Knutton, Longton, Fenton and Hanley, and later covered Wolstanton, Shelton, Clayton, Seabridge and Penkhull. In Penkhull, as we have already seen, the Manorial Court of Newcastle used to meet in the Greyhound. This rural manor was organised like any other with labourers paying rents and doing service in the fields, but the people in charge were in a very special position. They were called sergeants, and had all the baronial privileges of farming, running mills and collecting taxes, but they did not have to do any service outside the manor. They were not required to go off and fight in foreign wars in Wales and Scotland and France. Instead, they had to guard the castle. The names of some of these sergeants have descended in local family names like Michel, Ford and Parker, and the Burton family who monopolised the office for years.

The townspeople were used to monopolies and privileges and generally behaving like the chosen few. Perhaps that is why they are so upset when you confuse the Potteries and Newcastle. One of the more hilarious privileges was enjoyed by William de Brompton and his wife, Margery. John of Gaunt, Duke of Lancaster, virtual ruler of England and for certain King of Castille, was also Lord of the Manor of Newcastle-under-Lyme. In the 1370s he tried to sort out the finances of the manor, and one of the customs he discovered was that our William and Margery de Brompton, and all her ancestors, were accustomed to have, from time immemorial, four and a half silver pennies from each minstrel who came to the town to plunk his lute and moan his ditties at the feast of St Giles, to whom the parish church is dedicated. As if this was not enough, they were to get fourpence halfpenny from each bear that came to the borough to be baited. That really is too much. Picture those bears, queueing patiently with four whole pennies in their paws, and one half of a penny, cut down the stem of the cross on the 'tail', to pay for the privilege of being torn to pieces by Staffordshire terriers on the cobblestones of the market place.

That information is from John of Gaunt's Register, and it really does say "from every bear that comes to the Town to be baited, $4\frac{1}{2}d$". Those pennies that the poor bears paid with were made of silver from British mines, and were about the size of the recently defunct sixpenny piece, but not so thick. On one side thay had a portrait of the king, and a name which seems to have been *henricus*, whatever the king was called, and on the other side was a cross with fancy ends.

The cross eventually stretched from side to side to prevent people from cutting little bits off, and to make a halfpenny, you cut it in half, along one arm of the cross. Quarters, or 'farthings' (fourth-ings) were made by cutting the halves again, along the remaining arm of the cross, rather like cutting up a 'ring' of scones. I have several of these 'long cross pennies' which were supposed to have been found near the Castle site, but the story of their discovery is probably apocryphal. The coins show none of the usual signs of lengthy interment, and, surely, no local workman would be so naughty as to keep such a find, when his proper course of action would have been to surrender them to a coroner's jury for a Trial of Treasure Trove? Not much.

It is reasonable to ask what such small coins could buy, apart, that is, from the privilege of singing in the streets of Newcastle, or being torn to shreds there by the local hounds. The truth is that there was not much to buy in the usual sense. There was still a healthy system of barter in use, and people were sickeningly self-sufficient in the matter of shoes, clothes and food. There were specialists, of course, and a look at old records shews what some of the specialisations were. Felters, 'Taylours', Smiths and so forth catered for the helpless townsfolk rather than the country people, and the specialists could only accept so much payment in the form of produce and would have wanted some in the form of silver which was easily carried and easily hidden. Coins were used when trading became too unwieldy, and were also used to pay taxes and tithes, although parsons entitled to collect tithes amassed monstrous stores of produce, until quite recently, in barns like aircraft hangars. Memories of this unhappy system are aroused by that splendid Staffordshire Pottery figure group, the *Tithe Pig*, which shows a young man and woman in tattered clothes and a parson, dressed for duty on an oatmeal packet. Around the rector is a pile of farm produce, representing one tenth of all that the poor devils have produced, and the mother is proffering a child, presumably her tenth. No doubt the parson took the child and eventually sold it into one of the many forms of legalised slavery to which the Establishment subscribed at one time.

Apart from taxes, tithes and anvils, the silver pennies could purchase a most unusual commodity, freedom. In 1401, John Roe was elected as a freeman of the town, and paid for the privilege ten shillings, five shillings in hand and five shillings to be paid on Christmas Day, provided he had the backing of two guarantors. His mates, John Yorke and Henry Breton, sponsored him. This freedom was a valuable and much sought after thing. The freemen were excused from all forms of feudal service and from holding many of the offices that people found both expensive and irksome. They were able to get on with their lives in their own way and it was from this group of embryonic middle-class citizens that all the professional and technical activities sprang, the legal, medical, educational and domestic artificers who could only emerge from a self-governing community like Newcastle, who made the present state of the western world possible, for good or ill, and who are largely absent from the emerging countries now.

Many of a 'great family' of today, who would like to think they are descendants of Norman Conquerors, and are not, got started with the sale of a freedom to some scruffy, energetic, intelligent, lucky, just-about-grown-to-manhood back street urchin, who derived his name from the feudal estate on which he had previously 'serfed'.

Some of us could do with a few of those magical silver pennies with their hot-cross-bun backsides and their gormless frontside portraits of *henricus* whose wide-eyed innocence belies his nasty nature, whatever his real name was.

With the town set-up as it was, the market was of paramount importance. By 1173 Newcastle had grown so prosperous, and its townsfolk so numerous, that it was recognised as a borough. Sometime between then and 1179, *henricus* granted them a charter allowing them a number of liberties and free customs. This particular *henricus* was really called Henry. He was Henry II. We do not know the exact date of the granting of the first charter and only know of it second-hand from a charter granted to Preston in 1179, but it must have been a good one. The people of Preston were very careful to see that their charter was the same as Newcastle's, which they used for a model. They probably borrowed it when they had theirs drawn up. That might be why it is missing. They borrowed another charter from us in 1344 and kept it for over five hundred years; perhaps they have still got the older one, or maybe somebody has made it into a lampshade or something. Anyway, in 1973 the Mayor and townspeople of our Newcastle-under-Lyme took the longer view and celebrated their octocentenary, regardless.

When the proud, new Borough set out their town on the ridge, they made two very wide streets in the form of a huge 'T'. One is the High Street and the other the Ironmarket—a significant name. Both were intended as wide open spaces from the start, and the plan is still obvious.

The High Street starts by the big roundabout, with its sunken garden, on the A.34, and goes on up to the church. The whole of that area was originally clear, a huge open space for cattle pens and market booths. There were also streams running through it, but they are all culverted now. Their original market was on a Sunday, then it was moved to Saturday, and again changed to Monday, which has remained Market Day ever since.

The original markets must have been breathtakingly exciting and entertaining for the people who came. And they would have come from miles around, for Newcastle's was the only proper market in the area. They had the same sort of crowded, colourful bustle that North African markets have today. Merchants came from Stafford and other towns to sell their goods, and all paid a toll to the burgesses of the Borough. Every so often a visiting merchant overstepped his rights and the guildsmen and burgesses flexed their muscles, impounded his goods and sent him off with a flea in his ear.

Although the weekly markets were an economic necessity, they were also gay affairs, and gingerbread stalls, jugglers and soothsayers would appear from time to time, the great occasions were the fairs. Fairs were only held by special permission granted in a royal charter, for which the King, naturally, charged. But it was worth it; whilst the markets attracted people from near and far, the fairs attracted them from even further afield and it was to these riotous events that the minstrels and bears came. But not only those; jugglers, tumblers and acrobats, horses that could count, sheep with two heads, dwarfs and giants, fat men, bearded ladies and terrible Turks all turned up to entertain the multitudes. What a sight, with gay awnings on the booths, the colourful costumes and the beribboned stage-carts of the mummers. How they danced, and ate and drank, and fought. What stories the peasants of Burslem, and Handley Green and Normacot must have taken home to Grandma who was too old to walk the distance any longer. Did they really swallow swords and eat fire? Could our old dog be taught to walk about on his hind legs dressed in a vest like a Frenchman?

It is impossible to over-estimate the attractions of Newcastle Fairs to the medieval peasants of the surrounding villages and hamlets. Accustomed to a grey, listless existence of grinding poverty, malnutrition and ceaseless hard work, eked out in fetid hovels made of mud and straw, those fairs in the borough at Trinity, Easter and the Feast of St Giles the Abbot must have been a rare glimpse of something mercurial. No wonder they believed in fairies and demons. As an added attraction, gangs of rowdies frequented the fairs and markets stealing and breaking up the stalls and generally behaving in the time-honoured fashion for louts everywhere. In one particularly serious incident in 1320, an armed band of the Duke of Lancaster's men raided the town on market day, looted the stalls and set about the townspeople, leaving the High Street looking like a battlefield. The times were violent anyway, and to add to the general low state of civil standards, we were at war with Scotland, on and off, and the royal representatives were stirring up Wales with coldly calculated insolent heavy handed misrule.

There were complaints of assaults both individual and wholesale. One gang even broke into Heleigh Castle, the home of Joan, Lady Audley, and beat up her soldiers. They seem to have had it in for anyone concerned with the Audley family. In 1323, Thomas de Warwyk, who had been Joan Audley's clerk, publicly insulted Roger Swynnerton Jnr, beat him up and generally maltreated him. Not long afterwards, Roger's brother, Richard, with his gang including Richard, the son of Adam le Hirdeman, Richard de Childerplawe and William, son of William the Smith of Chelle, chose another market day to visit retribution on the luckless Thomas de Warwyk. They chased him round the stalls and, when they caught him, set about him with cudgels and beat and wounded him "almost to death".

The feud between the citizens of Newcastle and the poor folk at Heleigh did not stop there. Shortly after the attack on her previous clerk, Henry, Countess Audley's new clerk, together with his friends, Henry le Peleter and John de Iselwalle, came to Newcastle on yet another market day and, "like common malefactors" beat up Agnes, the wife of Robert del Bakhous [Bake-House] and also set upon and wounded Adam de Lanton, Jnr.

On an even more notable Newcastle market day in 1325, Adam Deneys, of Congleton, marched upon the town with an army of five hundred unknown men. The day was the Feast of St Gregory and, amongst other things, they took the goods of William de Snethe amounting to ten shillings worth of linen and woollen cloth, twenty shillings in silver, five quarters of oats valued at fifteen shillings and other foodstuffs amounting to twenty shillings, an appalling loss for one man. One of the gang was recognised as Ralph de Fouleshurst and he personally stole from Stephen Bonetable a brass pot worth ten shillings.

The thugs and hooligans who made all the fuss were not the only ones to benefit from the uproar they created. Riots were then, as now, a popular spectator sport, and the sellers of refreshments and no doubt the bookies as well, laying odds on the town against the outsiders, made a bit on the side. Not only this, the King, *Henricus*, knowing how disturbed times were, kept for himself all the fines of those brought to justice for their part in the weekly blood-letting bouts in Newcastle market place.

You only have to scratch the surface in Newcastle to find the bones of its long past just below. Every so often someone digs up the road and a wealth of types of road-making materials comes to light for a while. There are layers consisting of clay, stones, bricks, granite blocks, wooden blocks, concrete and tarmac. It pays to stop at a hole in the road, pass the time of day with the gentlemen leaning on the shovels, and admire the assorted strata that in some cases have lifted the surface of the road by a good two feet. Many of the shops in the Ironmarket used to have steps up to their street doors. Now that the road has swollen so much there are very few left with any step at all, and if things go on for long as they have, they may soon need a step down into the premises.

The Ironmarket is a fascinating street; like the High Street it is wider than usual, and was obviously made like that so that booths could be set up in the street, and so that goods might be displayed in the front of shops where the pavement is now. The shops probably sold nails. The hills around Newcastle are thick with iron ore, and the forests were dense and could provide ample charcoal for a smelting industry. At nearby Holditch the Romans were smelting iron nineteen hundred years ago, and ironworking was the borough's chief industry right through the Middle Ages, as the local names, Ferrer and Bloomer, show. There was also constant reference to ironworking in the town records from 1380 to 1840, and most of those mentioned were nail makers. As recently as 1825, Joseph Lovat was summonsed for erecting a nail factory in the town and "occasioning divers noisome and unwholesome smokes, smells, and stenches". In the 1861 census twenty-six people are named as being nail-makers, but the introduction of wire nails soon brought about their ruin. The nails they made would have been "cut-nails", that is tapering nails cut with a chisel from the diminishing edge of a sheet of iron; these metal chips were then placed in a clamp and the wider end bashed flat. This all pre-supposed a supply of iron plates, and we are indebted to our old friend, Doctor Plot, for information on one maker of flat iron plates.

He was John Holland, one of only two frying-pan makers in the whole of England. When you consider how the Englishman loves his bacon and eggs, Mr Holland and his anonymous colleague, of Wansworth in Surrey, must have had a prodigious trade. Doctor Plot does "not scruple to give the reader the full procefs". I have not the nerve to describe it in all of his detail, but it leaves us in no doubt as to the high quality of the iron used and the skill of the people involved in the manufacture. The first half of the job was carried out near Keele Hall, now part of the university. Local smelters provided bars of iron which were hammered into round flat plates at the forge set up specially for the purpose. He says that this differed little from those in other iron-works, having a hammer of about half a ton. The trick that gave them the edge on other manufacturers was in producing the plates in a heap, hammering several together at one time, thereby not only saving the fuel that would have been needed to heat each one separately, but also ensuring that the plates stayed hotter longer for clinging affectionately together. That was where the skill came in. The process he describes is exactly the one that turns little bits of iron into large lumps by welding, and the iron workers at Keele must have got the management of forge temperatures to a fine art to keep their pieces hot enough to work, and yet not so hot as to weld together. The interleaving of the iron plates had another beneficial effect. It cut down the amount of scale due to oxidising that wasted so much metal in the primitive, one-bit-at-a-time methods of Jacobean England.

The flat-work was then sent down the hill to Newcastle for further treatment. Once again the wily Mr Holland stacked them in little heaps, nine at a time, and nine at a time hammered them into nine different sizes of frying-pan with as many as twenty different sizes of hammer. The forge also turned out dripping-pans, but these were made singly and not plurally like nests of Russian dolls.

In the times of the Tudors and Stuarts the Borough abounded with smiths, ironmasters, ironfounders and nailers. The guild regulating the activities of these busy people met in an "yron hall" in the Ironmarket. Where else?

Going from the sublime to something else again, a companion industry during the seventeenth and eighteenth centuries was that of hatmaking. At that time the raw material for hats was felt, since they were felt hats, and large numbers of people in Newcastle were, as a consequence, felt-makers. This was a hard, skilled job involving acres of sheeps' wool and oceans of water combined with all manner of combs and mallets. As usual there was the question of quality. In felt it depended on the length of the hairs involved as well as their thickness. There were tedious hand methods employed to sort out the fibres, and they provided a living of a sort for a lot of people, but they kept the profits low for the master hat-makers. One of these, James Astley Hall, who was a Newcastle man, invented two machines for the materials side of the industry. One was a carding machine for separating the hairs, and the other was an ingenious wind machine that sorted the sheep from the goats.

The bulk of the hats they made were finished in London; Newcastle supplying the basic floppy shape, the smart bits being done in the metropolis. Because of this division in the hat-making process, it was more susceptible to changes in fashion than many other industries where the customers had to have what was available, rather than what the specifically wanted. What they wanted in the first half of the nineteenth century was, increasingly, silk hats for the rich and cloth caps for the poor. The hatting trade of the Borough diminished in proportion to the rise in popularity of the topper, and despite a spirited counter-attack by the producers of straw hats, by the end of Victoria's reign, the topper was supreme in prestige 'people-tiles' and Newcastle hatters faded away.

However, the businessmen of the borough were nothing if not diverse, and were pretty well as busy making clay pipes as they were making hats. The pipe-makers were well-established before the Civil War, with more than a dozen families involved in the work during the eighteenth century. One of them, Charles Riggs, is mentioned by the redoubtable Doctor Plot who was very impressed with the pipes he made and commends him as the originator of mass-production techniques. He says, "Charles Riggs of New-Castle, has a sort of Engin I never saw elsewhere, with which he punches the bolls of his Tobacco-pipes much quicker and truer than others of his trade, unacquainted with this instrument." Not that there is much to be seen of pipe-making now, since their wares have been replaced by other, more durable materials. There were two makers still at work in 1880: Thomas Broom at the Higherland, and George Lakin, in Penkhull Street, now the southern end of the High Street. However, the Museum, in the Brampton, has a good selection of pipes made locally, and, where almost any spadeful of garden soil in England will come up with one small piece of broken clay pipe, a spadeful in Newcastle will often result in a number of such pieces. Once, in a moment of youthful enthusiasm, I set a class of ten-year-olds to finding the 'bolls' of the tiny, seventeenth-century pipes. Within a week every window ledge and table-corner was littered with them. Only one had a maker's mark on it, and that was from Broseley.

Another occupation that kept the people of Newcastle busy was shoemaking. In 1880 there were fifty-two boot and shoe makers and seven cloggers, twice as many as in Longton which had a slightly larger population. Whether half the people of Longton went barefoot or bought their shoes in Newcastle I do not know, but it would accord well with the opinion that Newcastle people had of themselves to wonder if they, perhaps, had two pairs each? On the other hand, or foot, Hanley with nearly three times Newcastle's population had only one and a half times as many boot, shoe and clog makers. Those old, brown photographs of hollow-eyed school children wearing ragged jerseys and no shoes begin to make sense when we start to examine details like these.

Apart from being well-shod, Newcastle children also had the opportunity of a good education. One of the schools built in a fever of good works in the 1880s, the old Orme Boys' School, is still in existence. It is on the opposite side of Pooldam from the Queen Elizabeth Gardens, and had it been built a couple of hundred years earlier as its style hopefully suggests, it would have stood ankle deep in the Castle pond. When built it was named the Orme's English School, and was described as being "an Elizabethan stone erection in the Higherland". The name comes from the Reverend Edward Orme, a former headmaster of Newcastle High School, who, in 1704, left most of his money to found an 'English' school. He did this because the old Free School, started in 1602 by Sir Richard Clayton, to "teach and instruct in learning thirty poor children born and to be born within the said town of Newcastle-under-Lyme, gratis" had become a grammar school and no longer catered for poor children and was no longer "gratis". It had become the resort of the new privileged middle class who had begun to emerge like the misshapen grotesques of Tudor shopfront canopy supports when the feudal disaster of the Wars of the Roses gave way to the Welsh tyrants and the piracy and simony they encouraged. In the seventeenth century it took these people no time at all to forget their own humble origins, and turn upon the poor from whose ranks they had most lately emerged.

Where every other school charity and school came unstuck, Edward Orme's went from strength to strength. Whilst the other charities were carefully squandered by their trustees, who submitted the funds entrusted to them to all possible forms of legal plunder, Orme's funds were invested in land, including some at Knutton which eventually sprouted coal and iron mines like poppies after the desert rain. Apart from a period of about twenty years in the general confusion after Napoleon's wars, all went smoothly with Orme's gifts. By 1880, it had become the famous Newcastle Middle School, attended by Arnold Bennett and many other local successes. In 1875, the piece along Orme Road was built, at a cost of £1,500, and it was then able to cater for three hundred boys. It is a common error to suppose that what went on in school buildings when we were pupils is how it has ever been. Education has changed many times over the years, and the 'Orme Boys' has seen most of the greatest changes. According to a report of 1880, the course included "instruction in Latin, French, mathematics, geography, grammar, chemistry and drawing". Only one quarter of the scholars were educated free.

However, 'exhibitions' were created; one half of them were, in the first instance, to be competed for by boys who were being educated in the "public elementary schools" in the Newcastle school district, in order to give the more promising pupils a higher education, and in default of there being enough bright boys from the schools, they were to be thrown open to all comers.

The school year was divided into three terms, the fees were twenty-eight shillings per term, and, in 1880, the number of boys in the school was two hundred, of whom twenty were 'free exhibitioners' from the elementary schools.

The school was built on a site previously occupied by the workhouse and jail. To reach the jail from the town, people had to cross the Pool Dam, which was in the parish of Stoke, detached, because the castle and its pond, or where the pond and dam used to be, were owned by the Sneyd family whose seat of power had been Stoke. This derived from the period when the borough and manor of Newcastle were separate, and the latter included much of Stoke. The disadvantage of this situation was that the Borough Constables had no power when they were going from the corner of the Gardens to the corner of Orme Road. Miscreants, who had been convicted in the Town Hall, and who were being trudged off in chains, weeping and clanking, to the town brig, were sometimes 'rescued' by enterprising gentlemen who seized the prisoners from the clutches of the powerless constables, and either collected from the newly-liberated lags, or ransomed them to the Constables who were liable to both dismissal and financial sanctions if they lost their man.

The jail and workhouse were adjoining, a sort of semi-detached, plural hell. When more-thorough-than-usual maintenance was being done on the floor of the hall in the old school, workmen found a very deep well, covered only by a few rotting boards and a scrape of earth like a pauper's grave. Heaven only knows how many hundreds of reluctant, shuffling schoolboys have stood on that spot mumbling and murdering a perfectly good hymn, saved from a sooner-than-usual plummet to the depths by the kind of providence that looks out for the innocence of fools and the callowness of youth.

The tired old buildings are still useful. The school is now occupied as a Youth and Adult Centre, so that, perhaps, the Victorian dream of using it for "higher education" has come true at last.

Another complex of old school buildings that is now used as a Youth and Adult Centre is Ryecroft Schools. The schools were built in 1871 by the newly-formed School Board, and were then described as "an excellent range of schools, of brick with Bath stone dressings, for boys, girls and infants, giving accommodation to 800". Notice how infants in old schools never have any sex.

Motorists driving through the borough nowadays will have an opportunity of seeing those rambling old buildings with their Bath stone dressings and ornamental tiles, and their separate entrances for nasty little boys, sweet little girls and neuter infants. The brand new ring road, that now clasps our Old Lady Newcastle in a corset of concrete and reconstituted stone blocks, takes the traffic right across Rye Croft, where I doubt if even rye could survive in that concentration of tetraethyl lead and carbon monoxide.

Going back into the town from the old Orme Boys' School you will no longer be in any danger of being rescued by the local kidnappers, but you will have to cross the main London to Carlisle road. There is a crossing and a subway to help pedestrians over the busy tarmac. It is interesting to see that the modern planners came to the same conclusions as the Anglo-Saxons. For several hundred years the road ran through the town to provide customers for the shops and inns, and confusion and despair for old shopping-ladies. The old road had run along the edge of the castle pool and then climbed the hill through to Lower and Upper Green to escape to Congleton and the north.

Now that old shopping-ladies and young pavement-skipping children are considered by the planning-uncles, the main road has been re-routed along its old track, and, for what it is worth, anyone travelling along the A34 through Newcastle will be following much the same path as King John when he visited the Castle at the Ides of March, 1206. This king was not the stupid cowardly monarch of legend.

Then Newcastle-under-Lyme had a part to play in the desperate game of politics he intended to play, and he was here, carefully laying his plans nearly ten years before the event. John's irresponsible, lout of a brother, Richard, spent most of his time and our money at the Crusades or indulging in a little mild royal robbery on the continent.

Richard the Lionheart left a sort of a Viceroy to look after things for him in England. He was William de Long-champ who was thought to be more reliable than Prince

John, who might have expected to be left as Regent during his brother's absence. Richard was well aware of John's ambitions, and Longchamp was forced into open warfare to protect his absent master's interests. Several of the royal castles were reinforced to curb John. One of the most important and costly of these, together with its Loyal, though not too ancient Borough, was the New Castle Under Lyme. It controlled the main North road and its junction to all points west, and it commanded the southern border of the County Palatinate of Chester which was in favour of John. That is why he was here in 1206. He was feeling out the loyalties of his subjects as he could see trouble brewing amongst his ambitious barons. John had succeeded to the throne in 1199 after the Lion Heart had died from gangrene from an arrow in the shoulder. The arrow was shot during a kingly thieving expedition in France which was as disastrous as it was disgusting.

Six weeks after signing Magna Carta, in mid-June, 1215, an agreement he had no intention of keeping since it was made at knife-point, King John gave the New Castle that he had so painstakingly and expensively fortified nine years before, to Ranulph de Blundeville, Earl of Chester, knowing that the earl would help him in the impending civil war.

John got Papal absolution of his oath from Pope Innocent III, who died shortly afterwards, making the deceit irrevocable, and set out on his successful campaign against the Barons, who had grown too big for their tin-wellies, anyway. Apart from his English soldiers, who were, as ever, as always ready for a punch-up, John had a large army of Flemish mercenaries, some of whom may have found their way to Newcastle to reinforce the garrison there, since local legend has it that Rotterdam, between Knutton and Newcastle, was settled by Dutchmen as a reward for helping the Loyal and Ancient Borough in time of trouble. And no one had more trouble than King John.

All that from just crossing the road! Once these towns take a grip, like the local terriers, they never let go.

Supervising the busy traffic island is the parish church of St Giles the Abbot, the fourth to be erected on the site. Except for the tower, it was completely rebuilt in 1873-6 from the designs of Sir Gilbert Scott, R. A., at a cost of about £15,000. It was consecrated and opened on 27th May 1876. It is a spacious and lofty building, in the decorated style, consisting of a chancel and chancel aisles. There is a vestry to the north and the nave has a clerestory and aisles. There are porches at the north and south ends, and the old, square tower which dates from around the end of the twelfth century was not too badly handled, being restored at a slightly later date.

The aisles are divided from the nave by an arcade of six bays. With moulded bases and beautifully carved capitals, the pillars are alternately cylindrical and octagonal; above these the clerestory is lighted by six windows with alternate cinque-foils. There is a good deal of stained glass about the church, and some of it is very pretty.

The lectern, which is a survival from an earlier version of the church, deserves a close look. Where many such bookrests are made of brass in the form of eagles or vultures or whatever, the one in St Giles is in the form of a pelican with outspread wings, nourishing its young with blood drawn from its own breast. At the base is an incised legend "*Sic Christus Dilexit nos*", meaning 'Thus Christ Loved Us'.

There is a splendid Early-English arch opening from the tower to the nave. During some alterations to the church in 1848, the effigy of a man was discovered. Probably part of a tomb, he wears a long robe, in his left hand he holds a sword, and in the other a glove.

When the church was being demolished in 1873, a number of ancient tiles were found. From the design on these the new tiles for the floor were produced. The cost of making this superb link with the intentions of past church furnishers was the then staggering figure of £600, entirely met by Mr M. Hollins.

The Georgian church had been possessed of an apse at the Church Street, or East, end, and the change to the present shape represented both a return to an original plan and a radical rearrangement requiring extensive new foundations both to support the massive new stone walls, which were to replace the elegant brick apse of the previous structure, and also to prevent the church gliding off down the hill on to the main road.

During the digging of the new foundations the excavators predictably came across vestiges of the Early-English structure, and a stone coffin. The stone coffin was deposited within as a curiosity, and Sir Gilbert obligingly incorporated the disclosed elements of the original structure in his new design which is probably why the church looks older than it is. It is also an excellent example of how well the Victorian-Gothic restorers did their homework. The purpose of medieval ecclesiastical architecture was to create an atmosphere of sanctity combined with a feeling of light and space. Their nineteenth-century imitators have succeeded very well in St Giles. They used to have some splendid wrought-iron gates, a magnificent example of local iron-work, but they were pinched at the time of the restoration in the 1870s, and are supposed to be in one of the London museums. They all deny it, and we have to make do with a barrier that looks as though it was purloined from a Berlin Wall checkpoint.

Next door to the church is a small range of lovely old buildings, occupied in part by the premises of a chemist and photographer, and, at one time, by the Labour committee rooms. The chemists, who are leading lights in the Civic Society who do such good work preserving our nice buildings, had taken good care of their half which incorporates some very old timber, but the weighty deliberations of the local party have done nothing to improve their part and it is in danger of falling down.

Standing in front of these worthy old shops, and looking up at the tops of the buildings lining Red Lion Square and High Street, gives a wistful picture of a late-Georgian English townscape. But do not look down. The mess that modern shop-fitters can make of an old building is quite sickening. Excellence of mechanical design and skilled workmanship cannot make up for a total lack of sympathy for the character of the street. Thank goodness for the odd wine-shop and pub that do at least make the effort.

One of the most obvious buildings in the High Street is the Guildhall. It was built in 1713, and has not altered too much in general appearance. A description of it in 1880 says that it was a large, oblong building of brick, ornamented with stone pilasters, it had a clock tower, surmounted by a cupola in which was an illuminated clock with four dials, presented by James Astley Hall, Esq., of felt-machine fame, at a cost of £250. All that seems to have happened since is that the arches between the pilasters have been filled in with brick to make the old market space into usable rooms, and that Tram-Wire standards have come and gone. The iron plate that used to hold the electric cable for the trams is still fixed to the west side of the hall, and unless you are fortunate enough to catch someone digging up the road to reveal the old tram-lines, as they do from time to time, the plate is all you will see of the old street railway that has by slow stages evolved through the Potteries Motor Traction Company into the National Bus Company.

One of the endearing features of Newcastle is the thrice weekly market: Monday, Friday and Saturday. Known from time immemorial as the Stones because of its having long taken place on the cobbles of the High Street, it is one of the best markets to be found anywhere. The families of some of the traders have been shouting their wares in that spot since the time of Robin Hood. The flowers of the Easter market are a breathtaking sight; bank upon bank of golden blooms under the bright, striped awnings making a Dutch flower festival look like a worn out, tatty old peg-rug. Early arrivals hang about, diffidently, not wanting to be the one who buys the first few bunches and starts the process of inevitable tarnish that ends with masses of crumpled cardboard boxes, brown and sad, and a carpet of trampled greenery to deaden the joyful tread of stallholders' boots as they hustle off, choking with the delirious mirth of affluence, to the night-safe hoppers of the surrounding banks. All of which seem to have made a special effort to be housed in a building of charm or character, or both.

On one of the banks is a plaque indicating that on the site was the butcher's shop of Richard Harrison, four times mayor of the town, and the sorrowing father of Thomas Harrison who became a Major-General in the Parliamentary forces, was second in command to Oliver Cromwell, signed King Charles' death warrant, and was hanged, drawn and quartered in London. He was arrested in Merrial Street, having refused to go into exile in Holland because he knew he was right. Samuel Pepys was present at his execution and testifies to Harrison's extraordinary bravery as he was stretched until all his main joints were broken, disembowelled whilst completely conscious, and then mercifully beheaded. He was unfortunate in being a strong, healthy man. Weaklings who were subjected to the same barbarity usually died before the disembowelling bit. Poor Thomas lived through to the bitter end.

Two roads lead out of Newcastle northwards. The A34 goes out through Congleton, another ancient borough, and on to Carlisle and the Scottish border.

The other goes a winding route, up the Brampton and over the ridge to the Fowlea Valley, Burslem and Tunstall. On its way it passes another of Newcastle's churches, St George's built at the beginning of the last century, but looking as though it had been there, four square, for ever, with its crocketed pinnacles.

At the top of the hill, and on the right, are the borough's Pitfield House Art Gallery and near to it the museum which holds very good collections relating to the history of the area. Newcastle once was a flourishing provincial centre for the manufacture of clocks and these are well represented in the town's collections, as are the results of an exhaustive investigation of the site of one of the county's oldest china-works. This was situated on the opposite side of the A34 from St Giles Church, where the Sammi Belle's pub is now. The buildings, which are a good example of an early-eighteenth-century house, were part of the works, and archaeologists from the museum team excavated the yard, hovels and kilns of the factory, once owned by a Mr Samuel Bell, who was making very high quality ware there from 1724 to 1744. After his death a London potter, William Steers, started to make a very good kind of porcelain there and so Newcastle was the first of the North-Staffordshire towns to produce porcelain on a commercial scale. They were contemporary with the earliest London china works at Chelsea and Stratford-le-Bow.

Past the museum with its Crimean War cannon, the road leads across the Marsh, now a recreation ground, which is much influenced by the traffic of Wolstanton Colliery, which really does have the deepest shaft in England, sunk to a depth of 3,135 feet. Not that the colliery commands much attention since it is 'down bank' of the road. It is the graceful steeple of St Margaret's Church that catches the eye. The base of the tower is twelfth or thirteenth century, the rest of the fabric having been rebuilt in the approved, Gothic manner in 1858 and subsequent years. The churchyard has the gravestone of a glass-maker, Randull Bristall, who died in April 1668, and of his wife Ann who survived him by thirty-five years. I should love to know where in the area he made his "Broad Glasse".

Another tombstone carries one of the few direct accusations of murder to be seen on a sepulchral memorial. It is the stone of Sarah Smith, who died on 29th November 1763, and says:

> It was G–S B–W,
> That brought me to my end
> Dear parents mourn not for me,
> For God will stand my friend.
> With half a Pint of Poyson
> He came to visit me.
> Write this on my grave
> That all that read it may see.

She was aged twenty-one.

The churchyard is stuffed tight with the graves of famous potters and other local magnates. There were once a large number of earthenware memorials to potters in the churchyard, but when last I looked there were none to be seen. Some of them are reported to be in Liverpool Museum, of all places.

Inside there are many memorials to local big-wigs, quite a few being Sneyds. There is also an alabaster altar-tomb of the sixteenth century on which are carved the recumbent figures of Sir William Sneyd of Bradwell, and his lady, Dame Anne. Sir William is in plate armour and his head is resting on his helmet. Their fifteen children are carved on the front and the two end panels, all in beautiful clothes and a very pious attitude. This has to be anyone's favourite family group; they all look so alike. And the fact that the children all have their hands neatly folded in an attitude of prayer makes the absence of their mother's hands even more macabre. Local legend has it that she was carved without hands as she had lived without them for many years. The story says that the lady was very proud and haughty, and that she was so bone idle she would not even dress herself. She insisted on having every last thing done for her, and expected to be waited on, hand and foot. Her husband decided that he would try to cure her of her imperious behaviour, and, since bearing fifteen children did not do the trick, he ordered that she could either dress herself or have her hands cut off since they were obviously of no use to her, cluttering up the ends of her arms.

She was too lazy to agree to dressing herself, and he was too stupid and stubborn to change his mind, so her hands were cut off and there she lies without them to this day. This refutes the suggestion that you need hands to stop your arms from fraying at the ends. Dame Anne's have not. Her stumps are very neat. I tremble to think what the legend would have been had she lost her nose, like so many other statues.

Past the church, which will detain anyone who stops for far longer than they imagine, the road becomes a village street, with many cottage-type properties made into shops, and little alleys running away to right and left. The illusion fades as the roundabout at Porthill comes into view; another such dominated by a church, St Andrew's.

The road then runs down the hill to Longport and across the Potteries 'D' road. As yet another roundabout plants its foot across the road there is an unrivalled view of the Potteries as it was and is. Up the hill is Burslem with its high walled works and enigmatic chimneys; to the left is apparently open country; to the right and down the valley is dereliction, ancient and modern, Shelton Steel Works and Wolstanton Colliery, and the forlorn skeleton of a forgotten engine house.

Once across the road, and canal, and railway and up the hill again is Trubshaw Cross. Turn right for Burslem; keep right on for Tunstall. The road goes by Westport Lake Park, scene of a triumphant reclamation scheme that has gone wrong, because the concentration of nasty bacteria in the mud of its bottom makes it lethal to little paddling children and big boating boys alike.

VII

TUNSTALL AND THE HILLS

AFTER Porthill, Longport, Middleport and Westport, the next port of call is Tunstall, metropolis for the moorland villages and the most northerly of the Potteries Towns. It was, for a long time, part of the enormous parish of Wolstanton. In the 1770s Tunstall was described as "a mere street, or rather roadway, with only a few houses, probably not more than a score, scattered about it and the lanes leading to Chatterley and Red Street". And Mr Aikin, who considered it to be part of the country around Manchester, in 1795, said Tunstall was the pleasantest village in the Potteries.

The market place, now called Tower Square, is the centre of things. There was once a town hall in the middle of the square, but it has vanished, and its place has been taken by a clock tower, built of a rather incongruous yellowish brick in 1893. The second town hall, built in 1885, is at the east end of the square, and is an ornate, Renaissance-style building with a nine-bay front of rusticated stone. The upper storey is made of brick with the inevitable stone dressings, and there are lots of lovely, patterned, moulded tiles, in 'posh-bricks' under the dentil frieze. In the centre of the front is an oriel window obscured by a Star of David, no doubt a bit of Anglo-Hebrew solidarity. Below the Badge of Judah is the date MDCCCLXXXV, to the left of the date are the words "Peace, Happiness", whilst on the right "Truth, Justice". It would be pleasant to feel that all the citizens who have clogged and coughed their way past that inspired hope on their principal public building without ever noticing it, have nonetheless achieved that four-part ideal state.

At the rear of the hopeful-hall-of-the-town is the covered

174

market, a considerable benefit on a rainy day. Like the markets in the other towns its stall-holders all seem to be jolly folk who know everyone and have just the thing you have always wanted although you did not know it. That market is one of the most cheerful places I know, and what makes a visit even more rewarding is the huge, ornate, iron girders, perhaps a little overwrought, which either hold the roof up or the walls down; it is not obvious which. The beams are a stage-whispered reminder of the old industrial past of the tiny town, which was producing iron long before the Romans even suspected that our island existed. The production of iron went on through the Middle Ages and made the area, together with its neighbour Kidsgrove, across the border in the borough of Newcastle, one of the most important in the Midlands. The fortunes of great families, like the Audleys, were intimately bound up with the iron, and later the coal, too, of the top end of the Fowlea Valley. Although Tunstall had been there such a long time, she did not develop as a pottery town until comparatively late. Despite this, Tunstall looks and feels older than her five sister towns, probably because her main early-building phase was of the second generation of urban industrial housing and factory buildings. The other towns had built an earlier generation of houses and so forth, and these have had to be replaced more recently than the first phase of Tunstall. Because of this the five towns that developed earlier look newer than the Johnny-come-lately who contrives to play the role of elder statesman.

Once again, it is necessary to look up to get the message. In the commodious square opposite the rosy town hall, the shop fronts and offices look modern enough to those whose gaze is ever seeking sixpenny pieces on the pavement, but for those who straighten their backs and look the day in the face, the upper stories are much as they have been since they were built just after the Napoleonic Wars. When the square is illuminated by the four-o'clock-in-the-afternoon sun, which shafts straight down it, sighting along the chamois-leather-looking tower as though to aim it right through the front door of the town hall, the whole arena is stuffed with light and the old buildings look like dowagers in ankle-socks, hoping to be allowed to go to a children's party.

At the top, or west end, of Tower Square the Methodist New Connexion church, known as Mount Tabor Chapel, survives from 1824. Built of brick with a stone front, it is now a building society office and hair stylists up to its waist, but from there on up it is all chapel, with its Venetian window beneath a mock-Classical pedimented gable. This was a determined congregation who, before they built Mount Tabor, had met in a cottage, then a timber yard and the joiner's shop when the weather was wet. They had outgrown their church by the early 1850s and built themselves a new one in what was then Victoria Terrace, now named Lascelles Street; it has been represented as the worst slum street in the Potteries. This is an area where street names reflect the patriotic fervour for the royal family at the time they were re-named; as well as Lascelles, we have streets named Connaught, Princes and Harewood.

The last two lead down to St Mary's Church and school. Standing under the great steeple and looking up it is positively totemic in appearance. The church itself is like an enormous, ecclesiastic, brick-gothic railway station. Named the Church of Saint Mary the Virgin, it was completed in 1859. It has an aisled and clerestoried nave with a gallery at the west end. The chancel is jammed in between a tower on the north-east, a chapel on the south-east and a porch on the north. The style is Early English and it is made of Victoria-plum-coloured bricks. This is where the people of the Potteries really assert themselves architecturally. Where every other local brick building of any pretentions has stone dressings to put some H's into the accent, the Tunstall Church of the Virgin has dressings of real, honest-to-God blue bricks, and various Potteries-Gothic ornaments to match.

Going north from St Mary's you will hit Ladywell Road and then Roundwell Street. The names commemorate two of the three main wells of the town; the other was Sugar Well, so named because the water was always sweet and clear. Tunstall had no piped water until 1848, and her citizens had to make do with water from one of the three main wells if they had not a well or pump of their own. In a hot, dry spell the well water was soon used up, and people sat up by the wells all night waiting for the level to rise enough to get them a bucketful. Those were the days when every yard had a rain-tub for lovely soft water to wash in, but water to drink had to bē carried in a can, and, of course, the precious fluid had to be paid for. At the northern end of Roundwell Street, three streets run down the hill to join it, and they are named after three American Presidents, McKinley, Coolidge and Hoover; their top end is in America Street. Opposite the bottom end of Hoover Street is a superb public house, the Victoria Inn, a most beautiful example of local brick work, with a mosaic portrait of Queen Victoria (God Bless Her!) and a most attractive cameo-tile title proclaiming "Parkers Brewery", now extinct.

The bricks and tiles of Tunstall's old brewery still hang precariously together up in Lascelles Street. It seems like a huge, puce, hump-backed monster in that scruffy, narrow street. The architect must have been a visitor to Egypt, because he has taken the design for the front straight from the Hypostyle hall of the great temple of Thutmosis III at Karnak, and for a cupola has copied one of the little temples at Edfu.

With all this packed into their buildings it is small wonder than the Tunstallites considered themselves the cultural leaders of the Towns.

Part of the cultural heritage of the town is wrapped up in the Prince of Wales Theatre, built in 1863 at the junction of what are now Ladywell Street and Harewood Street. It was, in its time, known successively as the Theatre Royal and then St James's Hall. Occupying a site on what was, appropriately, Booth's Fields, the theatre was taken over by the Salvation Army as their Northern Potteries Citadel, presumably on the grounds that not only should the Black Fiend be deprived of the best tunes, but also of some of his best buildings. The previous abode of Satan was taken over as the barracks for the army of Christ in 1882, and so it remains to this day, resplendent in battleship-grey paint.

My next excursion into the intellectual history of 'Tunster' was the search for their earliest cinema. This was the property of Alderman George Barber, one of the Potteries greatest 'characters'; he was a plain man and shrewd beyond the imaginings of Solomon. He was also very gifted in earthy wit and was never at a loss for a 'wise-crack'. His progress through life may be readily gauged from the title of his autobiography, *From Workhouse to Lord Mayor*, a book which deserves a far wider audience than it currently has.

An orphan, George Barber suffered the fate worse than death reserved for the very poor. He was taken into the Bastille, as the Union Workhouse was called by the not-quite-so-poor people who hurried past, with their coat collars turned up, not daring to breathe, in case they 'caught' that extreme of poverty which seemed to afflict some families, like the nameless horror of an unspeakable disease.

Young George soon overcame his early difficulties and found employment first in the local chemical industry, (Chemical Lane still runs along the valley of the Fowlea Brook) and then in the mines. He was a devout Christian, an ardent lover of music, and widely read in a self-educated way. He was one of the Federation's most distinguished early servants. If I had to choose one man as epitomising the character of the Potteries, it would be Alderman Barber. Starting with no advantages and precious little encouragement, he looked around for the up-to-date thing the Potteries lacked, and found them well up to the mark industrially, but sadly lacking in modern entertainment.

The wonder of the age in the early twenties was, of course, the cinema. Lacking capital to establish a permanent film theatre, he toured the area with a 'fit-up' cinema, partly of surplus material from the recent war, and partly of his own contrivance, all loaded into an old army lorry. He eventually built himself the first real cinema in North Staffordshire, in Tunstall. Enquiring where it was, I was told that it was now a bingo hall, and that it was in Bolvey Yard. Kidding myself that I was a dab-hand with maps and plans, the search began. This was an unknown piece of the Potteries. In the process of teaching the local history of the area for a score of years, I had gone over every square inch of this fascinating fifty square miles of industrial archaeological Aladdin's Cave, and Bolvey Yard sounded too good to miss. But it had been missed. The challenge to find the elusive location of Barber's Palace, known to its habitués as the Flea Pit, was irresistible. This was the place where the proprietor, who had the local schools given a holiday occasionally so that he could enlighten them with religious films, had stopped the show to tell the mothers that their habit of allowing their babies to "pee on the floor" was causing a muttering amongst his orchestra. This was because the seats, naturally, were on a slope, and anything spilled, poured or jettisoned on to that floor found its own level in the orchestra pit. This was the day of the silent film, when a cinema had a small orchestra to enliven the proceedings. Mr Barber's Band was the family Kirkham, entire. And they did not think that every film should be accompanied by chamber music, or the 'Water Music', for that matter. Old George had reason to address his clients from time to time on other matters; for instance, he was deeply hurt that his Palace was called the Flea Pit, and prefaced many a performance with a brief curtain-lecture to the effect that the only fleas in his cinema were the ones brought in by the customers.

None of which brought me to the actual spot where it had all happened. Admitting defeat, and surrendering my geography certificate and map reader's badge, I asked my informant to take me and show me where the Palace was. "But," said I, when we got there, "this is the Boulevard." "That's right," said he, "the Bolvey Yard!"

In 1925, when King George V and Queen Mary visited the Towns to confer upon the federation the title and dignity of City, the Royals and the civic dignitaries were all at lunch after the ceremonies. Alderman Barber, as befitted his place in the pecking order of local politics, was seated near the Queen. A ham salad was to be consumed and Her Majesty, with a facility born of long practice, removed the fat from her meat and decorously deposited it on the side of the civic plate.

"Here, Your Majesty," said George Barber, "If thae dustna want that fat, I'll eat it." And suiting his action to his words, deftly removed the rejected delicacy to his own well-filled plate.

The workhouse dies hard.

Of all the illustrious sons of the clay, George Barber is the one I admire the most. He is typical of the determination that characterises local people. When Hugh Bourne and William Clowes were rejected by Methodist congregations because they thought that the new sect had become too organised and establishment minded, and reacted by holding 'camp-meetings', they held a crusade for a more original and primitive form of worship. Their search for seclusion and peace to hold their services as they thought fit took them to Mow Cop, a couple of miles north of Tunstall and as wild and isolated a place then as you could have wished for.

Mow Cop still has much of that ruggedness and close-to-the-sky feeling that took the 'Prims' there for their first meeting. Preaching in lanes, sheds and yards, the two home-grown Potteries evangelists gathered support for their ways all over North Staffordshire. It was at Tunstall that the first group of revivalists met, in March 1808, in Joseph Smith's kitchen. Later on they built themselves a chapel at the corner of Calver Street and what is now Oldcourt Street. Once the 'Cathedral' of Primitive Methodism it has succumbed to the healing of old wounds, which is a good thing, and to the bulldozer, which is not always such a good thing.

The moors from which the infant Pottery industry was dragged, protesting feebly, into the affairs of the world in the 1700s, have always seemed nearer to Tunstall than to any other of the Six Towns, which, like any other big industrial town, has spread, fungus like, to engulf its nearer neighbours. There is not much indication, now, of the boundary between the Towns and the country villages, and the transition from Tunstall to her cousin, Biddulph, is hardly noticed on the journey by road. But Biddulph, which now acts as a dormitory for the 'top end' of the Towns as well as fulfilling her own function and duty to her townsfolk, is an extremely ancient settlement, perched on the edge of Biddulph Moor. Overlooking the beautiful Rudyard Lake, the Moor is as wild and splendid a stretch of countryside as you will find anywhere in Britain. Difficult for even the locals to take, legend has it that the inhabitants of Biddulph Moor are descended from Saracens brought back from the crusades to see if their experience in the Judean Hills would help in the taming of this rocky outpost. One thing is certain: some of the families have members who are darkly handsome. Recent archaeological discoveries have led us to suppose that these dark-haired very independent people are, in fact, a survival of the very ancient Brigantian people. Both alternatives are so romantically attractive, I do not know which one to choose.

These moors, which are only a cock-stride from the busy banks, are much visited by the people of the Potteries and are a solemn reminder of the closeness of the wilderness, both in time and in distance.

The smoke and grime are gone. The Dutch visitor who asked where we obtained the beautiful black stone for our churches and town halls, would now ignore the anonymity of the faded, tobacco-coloured relics that seem to have taken their place. As the *Sentinel* put it in a recent cartoon, "If God had meant them to be light-brown, He'd have made 'em light brown".

The bottle-ovens are now tourist attractions, no longer smog-generators. Gladstone Pottery Museum propose to 'fire-up' one of their handsome kilns, both to shew the new generation what they are missing, and to preserve the knowledge of the process before it too succumbs to progress and the Grim Reaper. No town on earth has been more abused by its own creation than has Longton by its bottle-ovens, yet the suggestion to set one of the fiery-bellied monsters loose again has met with eager approval from those who were most abused by them. It is as though they feel that the firing, which may assume the role of a once-yearly cult activity, will open a narrow window giving a backwards glimpse of a golden age. Even though they know that it really will be 'as through a glass darkly'. I wonder how they will be able to afford to buy the coal?

Our pit-heaps are now grassy knolls and the railway lines paths of bird's-eye and willow herb. Our tortuous streets that manfully refused to go anywhere are over-stridden by a great 'Queensway' that gives us the link with the rest of the country that the 'Knotty' was too coy to enforce. It is now very easy to get to the Six Towns, and, as many of the youngsters are finding, very easy to get out.

Such slums as there were are now banished to make way for neat chalets. But do not let the cleanness of newness fool you; the old Potteries is still there, in its new suit. Its people have not changed; like George Barber, they will still know how to rise from poverty to affluence and influence by taking hold of a modern invention at the right time, and by not compromising their standards, as Browning says in 'Rabbi ben Ezra':

"Time's wheel runs back or stops:
   Potter and Clay endure."